Why the Universe Exists

Why the Universe Exists

How particle physics unlocks the secrets of everything

NEW SCIENTIST

First published in Great Britain by John Murray Learning in 2017
An imprint of John Murray Press
A division of Hodder & Stoughton Ltd,
An Hachette UK company

This paperback edition published in 2022

A CIP catalogue record for this title is available from the British Library

B format ISBN 9781529381931
eBook ISBN 9781473629691

Typeset by KnowledgeWorks Global Ltd.

Printed and bound in Great Britain by Clays Ltd, Elcograf S.p.A.

John Murray Press policy is to use papers that are natural, renewable and recyclable
products and made from wood grown in sustainable forests. The logging and
manufacturing processes are expected to conform to the environmental regulations of
the country of origin.

John Murray Press
Carmelite House
50 Victoria Embankment
London EC4Y 0DZ

Nicholas Brealey Publishing
Hachette Book Group
Market Place, Center 53, State Street
Boston, MA 02109, USA

instantexpert.johnmurraylearning.com/

Contents

Series introduction

New Scientist's *Instant Expert* books shine light on the subjects that we all wish we knew more about: topics that challenge, engage enquiring minds and open up a deeper understanding of the world around us. *Instant Expert* books are definitive and accessible entry points for curious readers who want to know how things work and why. Look out for the other titles in the series:

Contributors

Editor: Stephen Battersby is a physics writer and consultant for *New Scientist*.

Series editor: Alison George is *Instant Expert* editor for *New Scientist*.

Articles in this book are based on talks at the 2016 *New Scientist* masterclass 'Mysteries of particle physics' and articles previously published in *New Scientist*.

Academic contributors

Jon Butterworth is a professor of physics at University College London, and a member of the ATLAS collaboration at CERN's Large Hadron Collider, who researches the mechanism of electroweak symmetry breaking, which explains why some things have mass. He wrote 'Why do we need the Higgs?' in Chapter 2, and Detector story plus 'The big discovery' in Chapter 3.

Michael Duff is Emeritus Professor of Theoretical Physics at Imperial College London, and a pioneer of supergravity. He wrote about string sounds in Chapter 9.

Dave Goldberg is a professor of physics at the Drexel University in Philadelphia, Pennsylvania, specializing in theoretical cosmology. He wrote 'Why symmetry rules the universe' in Chapter 2.

Andrew Harrison is CEO of the Diamond Light Source at Harwell, UK, and a visiting professor of chemistry at the University of Manchester. in the UK. He wrote 'Neutrons at work' in Chapter 11.

Eugene Lim is a theoretical cosmologist at King's College London. His interests range from string theory to the role of quantum information in the cosmos. He wrote 'Still no theory of everything' in Chapter 9.

Phil Walker is professor of nuclear physics at the University of Surrey in Guildford, UK, whose research focuses on nuclear isomers. He co-wrote 'Inside the atom' in Chapter 1.

Tom **Whyntie** was public engagement fellow at the school of physics and astronomy, Queen Mary University of London, having worked on the searches for dark matter and magnetic monopoles at CERN's Large Hadron Collider. He is now a research associate at University College London. He wrote 'What has particle physics ever done for us?' in Chapter 11.

Thanks also to the following writers and editors:

Robert Adler, Gilead Amit, Anil Ananthaswamy, Jacob Aron, Stephen Battersby, Michael Brookes, Jon Cartwright, Matthew Chalmers, Stuart Clark, Amanda Gefter, Jessica Griggs, Lisa Grossman, Joshua Howgego, Hannah Joshua, Valerie Jamieson, Kirstin Kidd, Elizabeth Landau, Christine Sutton, Richard Webb and Jon White.

Introduction

Since you opened this book, hundreds of billions of ghostly particles called neutrinos have passed through you. Protons with energies far greater than anything we can make in an accelerator have crashed into the upper atmosphere and produced vast showers of exotic offspring. And countless massive particles have lived a brief existence and disappeared again just to stop your body from flying apart at the speed of light.

The fact we know all this is testament to the ingenuity of today's physicists, who have revealed so much about the subatomic world. They have advanced our theories of matter and the forces that rule it. They have designed and built the instruments to gaze into the heart of matter, and worked out how to decipher the complex and sometimes subtle signals that those instruments are telling us.

This Instant Expert guide will take you into the realm of the particle. It will delve deep into the Earth's crust, zoom out into the cosmos, and travel back in time to just after the Big Bang.

The aim of particle physics is to understand how things work on the most basic level. What are the fundamental building blocks of everything in the universe? How do these elemental entities bind together to form more complex matter and how do they exert the forces we feel? This is a stunningly ambitious project, yet particle physics is also very simple. It consists of hitting things very hard, to find out what is inside them and how they work. Witness the Large Hadron Collider (LHC) – the most powerful thing-hitter yet devised by humankind, capable of reaching higher energies, probing finer scales and creating more massive particles than ever before.

Today the LHC has capped off a great theoretical edifice called the standard model of particle physics, which brings together all the known particles of matter and describes how they transform and interact with one another via a few fundamental forces, which are carried by another small set of particles. We now have a profound understanding of the workings of matter, based on mathematical symmetries and confirmed by huge experiments.

As this closes one chapter in fundamental physics, it leaves us wondering what happens next. Many things are missing from the standard model. From the shifty behaviour of neutrinos to the nature of dark matter, a compendium of particle puzzles is waiting to be solved. Will they lead us to some kind of final insight, or just another, deeper set of questions?

Stephen Battersby, editor

I
Fantastic particles and where to find them

Since we began to delve within the atom more than a century ago, we have found that our world is built from an array of objects with very peculiar properties.

Inside the atom

The idea of atoms as the ultimate indivisible particles of matter dates back to the philosophers of ancient Greece. It is the bedrock on which the new science of chemistry was built from the eighteenth century onwards. But all that changed more than a century ago, with glimpses of smaller, more basic entities – the first hints of what we now call elementary particles.

In 1897 the British physicist J. J. Thomson (see Figure 1.1) was investigating cathode rays – streams of radiation given off by metal electrodes under high voltage in a vacuum. These rays were invisible but would create a glow when they hit a fluorescent material. Thomson showed that they were bent by magnetic and electric fields, always by the same amount no matter what metal made up the cathode. He concluded that they were tiny negatively charged bodies, much smaller and lighter than atoms. The discovery of these 'electrons' put paid to the idea that the atom was uniform and indivisible.

FIGURE 1.1 J. J. Thomson (1856–1940), who discovered the electron – the first subatomic particle

If the electron were part of the atom, what else might be in there? To maintain the atom's overall electrical neutrality, Thomson suggested that electrons were embedded inside it, like plums in a 'pudding' of positive charge. But, by 1908, New Zealander Ernest Rutherford (see Figure 1.2), working with his assistant Hans Geiger at the University of Manchester, UK, had revealed a different picture. When fired from a radioactive source, positively charged alpha particles – later revealed to be the atomic nuclei of helium – passed through metallic foils placed in their way, deflected by just a few degrees. The atom, it seemed, incorporated a large amount of empty space.

Follow-up experiments by Geiger and his student Ernest Marsden delivered an even greater surprise. Some alpha particles bounced straight back, turned by up to 180 degrees. It was, as Rutherford later said, 'as if you fired a 15-inch shell at a piece of tissue paper and it came back and hit you'. Rutherford's interpretation, first delivered publicly in February 1911, was that the mass of the atom, itself less than a billionth of a metre

FIGURE 1.2 Ernest Rutherford (1871–1937), often cited as the father of nuclear physics

$(10^{-9}$ m) across, was concentrated in a tiny central volume just 10^{-14} m across. That is something akin to a fly buzzing around inside a cathedral – except that the fly accounts for 99.9 per cent of the cathedral's mass. The atomic nucleus was born.

Into the nucleus

For some time after the nucleus was discovered, its basic structure remained a puzzle. But when physicists transmuted one element into another using alpha particles, they found that hydrogen nuclei were emitted. By the early 1920s Rutherford and others were convinced that the hydrogen nucleus, later called the proton, must be a fundamental component of the nucleus. Only in 1932, though, did Rutherford's colleague James Chadwick isolate the other component. Bombarding beryllium with alpha particles produced a new type of radiation, particles with no electric charge (see Figure 1.3). At first, Chadwick thought it was a combination of electron and proton, but it turned out to be slightly too heavy. The proton weighs in at 938.3 megaelectronvolts (MeV), more than 1,800 times the electron's mass. The new neutron, meanwhile, tips the scales at 939.6 MeV.

While a proton left on its own is stable, or at least has never been observed to decay, a neutron can change into a proton by emitting an electron. If you could gather a bucketful of neutrons, after ten minutes only half of them would be left. Combine this with the fact that protons repel each other because of their positive charges, and it seems a miracle that nuclei stay together at all. That they do is down to the trumping effect of the strong nuclear force, which binds together protons and neutrons over very small distances (see Chapter 2).

With the electron, the proton and the neutron, we might seem to have a set of particles that could form any atom, accounting for all the chemical elements and therefore all known matter.

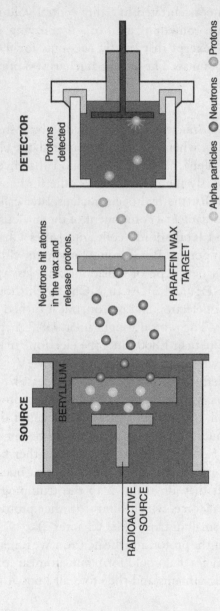

FIGURE 1.3 How the neutron was discovered. When bombarding beryllium with alpha rays, the French physicists Irène and Frédéric Joliot-Curie discovered a mysterious type of radiation capable of knocking protons out of paraffin wax. It was later discovered that neutrons were responsible.

And the newly developed theory of quantum mechanics captured the peculiar behaviour of these particles, which could act like waves as well as like little points of mass. Needing only three pieces to build the universe would be a stunningly simple system ... but nature did not turn out to be so kind.

Uncovering the anti-world

In 1928 English physicist Paul Dirac had already predicted a new type of particle. He devised a quantum equation for the electron that, unlike conventional quantum mechanics, also conformed to Einstein's special theory of relativity which describes how very fast-moving things act. This equation predicted that electrons have spin – a built-in ration of angular momentum. (The electron has a spin of ½ – in terms of quantum physicists' favourite constant, the reduced Planck constant, which is about 10^{-34} joule seconds.) It also showed that they should have a doppelgänger – an 'antimatter' particle with almost all the same properties but with a positive rather than negative electric charge. The positron was discovered in 1932, and it was only the first member of the anti-world to be revealed. There are also anti-protons, and antimatter versions of other particles too.

A still stranger side of nature was revealed at about the same time. When a neutron decays into a proton and an electron (one example of a process called beta decay), the energy of the two new particles adds up to less than the total energy the neutron started with. This shortfall led physicists Wolfgang Pauli and Enrico Fermi to conclude in the 1930s that a second particle must also be emitted – a ghostly, weakly interacting particle now known as a neutrino (specifically in this case an electron antineutrino).

More surprises were in store among the particles raining down on us from space, known as cosmic rays. In 1937 a particle about 200 times the mass of the electron was found among cosmic rays. At first, this seemed to fit a theory devised by Japanese physicist Hideiki Yukawa in 1933, which showed how new particles he called mesons might exert the strong nuclear force that binds protons and neutrons together. Instead, physicists found out in the 1940s that this new discovery was a heavier version of the electron. It prompted US physicist Isidor Isaac Rabi to say, 'Who ordered that?'

What was the purpose of all these extra particles? And was there another layer to discover? To dive deeper into matter and answer these questions, physicists would need to develop some powerful new tools.

Particle smashing

Particles are little things but, in order to study them, we need huge machines. The world's biggest and best-known machine for studying them is the Large Hadron Collider at the CERN laboratory in Switzerland. The LHC's tunnels are 27 kilometres long and, at peak consumption, this particle accelerator uses about 200 megawatts of power, around a third of what's needed to power the nearby city of Geneva.

The purpose of the LHC and other big accelerators is to take charged particles and boost them close to the speed of light. This gives the particles high kinetic energy and therefore a powerful punch. When such a high-energy particle collides with something else, its energy can be converted into the mass-energy of new particles (according to Einstein's equation $E = mc^2$). More energy means that you can create heavier new types of particle. This also enables physicists to probe matter at very small scales, because beams of subatomic particles act like waves. The higher their energy the shorter their wavelength will be; and the shorter the wavelength the smaller the object that can be discerned.

One simple particle accelerator is the cathode-ray tube – the device that J. J. Thomson was using when he discovered the electron. A glass tube, with the air pumped out, has an electrode inserted at each end. Applying a voltage sets up an electric field between the electrodes, and the negative end (the cathode) is heated so that electrons effectively boil off it. The electrons are then attracted to the positive electrode, gaining energy as they move through the electric field in between.

If the voltage difference between the electrodes is one volt, electrons will gain one electronvolt (eV) of energy, equal to about 1.6×10^{-19} joules. Turn up the voltage and you give the electrons more energy. Some cathode-ray tubes used for generating X-rays operate at hundreds of kiloelectronvolts (keV).

Higher voltages can be created in Van de Graaff generators, which use a belt to carry charge up into a metal sphere. They can reach millions of volts, and so can provide beams of protons at energies of several megaelectronvolts (MeV). That is high enough to probe the structure of the atomic nucleus, but still not enough for particle physicists.

High kicks

There are limits to how great a voltage can be sustained so, to reach much higher particle energies, accelerators make repeated use of smaller electric fields. In 1928 Norwegian engineer Rolf Widerøe built the first such machine, a type of linear accelerator, or linac. His linacs sent the beam through successive regions of alternating electric fields, phased to give particles repeated kicks as they travelled along. Some modern machines work the same way, while other linacs accelerate particles with a travelling electromagnetic wave, like surfers riding an ocean wave.

The biggest linac in the world is at the SLAC National Accelerator Laboratory in California. This machine is 3 kilometres long, and when it was being used for particle physics it would boost electrons up to 50 billion electronvolts (50 gigaelectron-volts, or GeV). Today it operates in two sections and its beams are used in other areas of science (see Chapter 11).

Linacs have their limits because particles soon reach the end of the line and you cannot accelerate them any more. This is why today's most powerful accelerators, called synchrotrons, use magnets to bend the particle beam into a circle. As the particles are accelerated, the magnetic field is increased and the frequency of the applied electric field is ramped up to keep pace.

The magnets must also focus the beam, or else particles would stray and hit the walls. In larger synchrotrons, long dipole (two-pole) magnets bend the particles, while quadrupole (four-pole) magnets focus the particle beam.

The modern synchrotron consists of an injector (usually a linac), a ring of dipole and quadrupole magnets, the beam pipe (with pumps to keep it in a state of high vacuum), and several radio-frequency cavities. These are hollow metal structures in which electromagnetic standing waves form, providing the electric fields that accelerate the particle beam.

Synchrotrons range in size from a metre or so across (used as X-ray sources) to the biggest in the world today, the Large Hadron Collider, with its circumference of 27 kilometres. The LHC occupies a tunnel originally carved out for an earlier machine, an electron synchrotron called LEP. This was built to curve as gently as possible, because when a high-energy charged particle moves on a curved path it emits radiation, called synchrotron radiation, and so loses energy. The amount a particle radiates increases the more it bends, as its velocity approaches the speed of light. Heavier particles move a little more slowly at any given energy, so protons can be accelerated to higher energies before synchrotron radiation saps their strength. For example, the LHC can accelerate its protons up to about 7 TeV, about 70 times the electron energy reached in LEP.

Once the particles are up to speed, they can be brought to bear. In some accelerators, particle beams hit a solid target, but it is much more efficient to collide two beams head on, as in the LHC. The collisions create a profusion of new particles, and huge particle detectors track this debris so that physicists can reconstruct what happened at the moment of collision.

Particle detectors

Some detectors simply count particles. Others measure the energy the particles lose. The most useful reveal particle tracks, as the paths of aeroplanes become visible through vapour trails in the sky. With the addition of a magnetic field to bend the paths of the charged particles, a tracking detector also provides information on their charge and momentum. Neutral particles are usually detected though the charged particles they set in motion when they interact in a detector.

The bubble chamber is one of the most famous types of detector. When a charged particle passes through the superheated liquid in the chamber it ionizes atoms, triggering the formation of tiny bubbles along its path. Many of the familiar images of particle physics are photographs from bubble chambers of the 1970s. Today images of collisions at the LHC (see Figure 1.4) are created from the electrical signals produced in huge instruments consisting of many layers of detector, each of which has a specific purpose to aid in tracking and identifying the myriad particles produced (see Chapter 3).

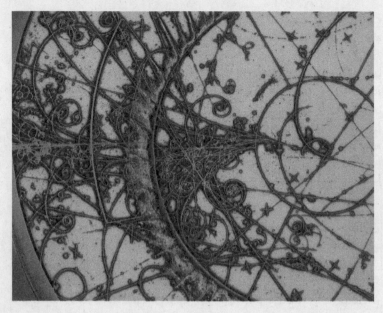

FIGURE 1.4 A bubble-chamber collision. This artistically enhanced image of real particle tracks was produced in the Big European Bubble Chamber (BEBC), which is filled with liquid hydrogen. Bubbles form along the paths of the particles as a piston expands the medium. A magnetic field causes the particles to travel in spirals, allowing charge and momentum to be measured.

Parts of particles

From the late 1940s, physicists started to find one new particle after another, many produced by cosmic rays colliding with atomic nuclei high in the atmosphere. Studies of cosmic-ray by-products revealed the first evidence for the pion, the kaon and the lambda, which are highly unstable with lifetimes in

FIGURE 1.5 A sculpture outside the visitor centre at Sellafield nuclear
power station in Cumbria, UK, shows a classical view of the atom
with electrons on well-defined orbits around the nucleus. According to
quantum mechanics, they are better described as diffuse probability clouds.

the region of 10^{-8} to 10^{10} seconds. Then came delta and sigma
particles, and more followed – more than a hundred seemingly
fundamental new particles, all of them unstable. Most of these
were fairly heavy particles, which along with the proton and
neutron were collectively called hadrons. In their search for
the simple basic building blocks of matter, particle physicists
seemed instead to have found a new subatomic realm of surprising and confusing complexity (see Figure 1.5).

Using particle accelerators to mimic the collisions of cosmic
rays under controlled conditions, physicists could make a more
systematic study of these particles. This revealed a new property,
which labels some hadrons as different from others and has no
analogue in the macroscopic world. Because this property leads
to behaviour that seemed strange, the property itself was called
strangeness. Of the particles mentioned so far, the proton and the
neutron have no strangeness, nor do the pion or delta particles.
The kaon, the lambda and the sigma particles all have one unit.

In the early 1960s the American Murray Gell-Mann and the Israeli Yuval Ne'eman independently worked on classifying the known hadrons according to their charge, strangeness and spin (a particle's intrinsic angular momentum). They found patterns of eight particles (octets) and ten particles (decuplets), which reflected a type of mathematical symmetry known as SU(3).

Three units of strangeness

One gap in these patterns corresponded to a particle with negative charge and three units of strangeness. Physicists called it the omega-minus, and in 1964 a research group using a particle accelerator at the Brookhaven National Laboratory in New York found it – a short but distinctive track in their bubble chamber. This showed that the theory had some predictive power. But what was behind these pretty patterns?

The mathematics of SU(3) shows that the larger groups – the octets and decuplets – are all built from a basic group of only three members. Perhaps the hadrons were based on an underlying group of three particles? Gell-Mann and, independently, another American, George Zweig, proposed that hadrons are indeed built from such basic entities. Zweig called them 'aces' but the name we use today came from Gell-Mann, who apparently liked the sound of the word 'quark' in a passage from James Joyce's novel *Finnegans Wake*.

They needed three types, or flavours, of quark, called up (u), down (d) and strange (s). As with all charged particles, there are also antiquarks of each flavour that have the opposite electric charge. By grouping quarks together in threes, we can build baryons – that is, hadrons with spin 1/2 (such as the proton, which is duu, or the neutron, which is ddu, or the lambda, which is dus) or with spin 3/2 (such as the omega-minus, which is sss).

Alternatively, we can combine a quark with an antiquark (with exactly opposite values of charge and strangeness) to make hadrons with spin 0 or 1, which are called mesons. These include the charged pion (u quark and d antiquark, or vice versa) and the charged kaon (u quark and s antiquark, or vice versa).

For real?

The idea of quarks was difficult to accept because of their fractional electric charges. In the nineteenth century Michael Faraday had established that electric charge always exists in multiples of some unit of charge, and J. J. Thomson's discovery of the electron in 1897 suggested that this was none other than the charge of the electron. But the new particles broke established rules by having charges of +2/3 or -1/3 that of the electron. This seemed revolutionary, and led many physicists to wonder whether quarks were artefacts of the mathematics rather than any real kind of particle.

But the reality of quarks soon got support from experiments (see Figure 1.6). There was already evidence that protons and neutrons were not simple spherical or point-like objects, because of the complicated way that electrons bounce off them. In the late 1960s physicists in California probed deeper: they aimed a beam of electrons from SLAC's 3-kilometre-long linac at a target of liquid hydrogen, and measured the energies and directions of the scattered electrons to put together a picture of what the protons looked like. The electrons were seeing tiny, point-like concentrations of charge inside each proton: evidence that the proton does indeed contain smaller parts. Finally, in the early 1970s researchers at CERN in Geneva confirmed that these parts carry charge of $-1/3e$ and $2/3e$, as the theorists had claimed.

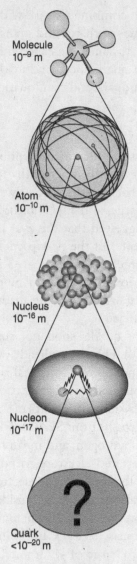

Molecule
10^{-9} m

Atom
10^{-10} m

Nucleus
10^{-16} m

Nucleon
10^{-17} m

Quark
$<10^{-20}$ m

FIGURE 1.6 Russian dolls: the particles that make up matter

In 1974 experiments studying electron–positron collisions discovered evidence for a new, heavier, fourth type of quark, which became known as the charm quark (its existence worked 'like a charm' to cure certain theoretical problems). A fifth, still heavier type of quark, called the bottom or beauty quark, showed up in 1977 in an experiment at Fermilab in Illinois. Here, the experimenters were studying muon–antimuon pairs produced in the collisions of high-energy protons with a target. This time they found evidence for a new particle some ten times heavier than the proton, which could be interpreted as a new heavy quark bound with its antiquark. Finally, in 1995 Fermilab researchers found a sixth, the top (or truth) quark.

Along with the strange quark, each of these three new flavours seems to carry its own peculiar property. For example, 'charmed' mesons exist, which contain a charm quark together with an antiquark of another variety. Quarks can change from one variety to another, and the top, bottom, charm and strange quarks all rapidly decay to the up and down quarks of everyday matter.

A small history of particle physics

5th century BCE
Greek philosophers Leucippus and Democritus speculate that matter is made of indivisible particles they call atoms.

1897
J. J. Thomson discovers the electron, the first elementary particle to be identified.

1905
Einstein proposes that light consists of discrete particles of energy, later called photons.

1947
The pion is found in the by-products of cosmic-ray collisions. This is the first example of a strong-force-carrying meson.

1937
A particle of 200 electron masses is discovered in cosmic rays. It was later found to be a heavy version of the electron – a muon.

1935
Hideiki Yukawa proposes a theory in which middleweight particles, mesons, carry a strong force between protons and neutrons.

late 1940s–1950s
Physicists discover an alphabet of new particles, including kaons, the neutral pion, delta particles and the sigma.

1957–9
Julian Schwinger, Sidney Bludman and Sheldon Glashow all publish theories of the weak nuclear force being carried by heavy particles; their work is based on the earlier 'gauge theory'.

2012
The Higgs boson is discovered.

1995
The sixth and final flavour of quark is detected: the top quark.

1908–11
Ernest Rutherford's team discover the atomic nucleus.

early 1920s
Rutherford and others realize that the hydrogen nucleus – the proton – is a component of all atomic nuclei.

1923
Arthur Compton confirms that photons act like particles.

1932
The positron is discovered. James Chadwick discovers the second component of the atoic nucleus, the neutron.

1930
Wolfgang Pauli suggests that apparently missing energy in beta decay is being carried away in the form of a new particle, the neutrino.

1928
Paul Dirac predicts that the electron should have a positively charged double: the positron.

1964
Murray Gell-Mann and George Zweig independently propose a fundamental particle, which Gell-Mann calls the quark.

1964
Peter Higgs is the first to explicitly predict the mass-giving boson – the particle that would eventually acquire his name. Robert Brout, François Englert and others produce similar ideas.

1983
The W and Z bosons are detected.

1976
The tau lepton, an elementary particle similar to the electron and muon but heavier, is discovered.

1973
The theory of quantum chromodynamics describes the strong force in terms of quarks, gluons and colour charge.

1967
Steven Weinberg and Abdus Salam combine electromagnetism and the weak force into one 'electroweak' interaction.

Why are particle masses given in units of energy?

As Einstein's theory of relativity showed us, mass and energy are linked by the equation $E = mc^2$. You can make a unit of mass by taking a unit of energy, say the joule, and dividing it by c^2 (where c is the speed of light). In particle physics, a convenient unit of energy is the electron-volt (eV) – the energy gained by an electron (or proton) moving through a voltage of one volt. Particle physicists use units of thousands, millions or billions of electron volts (keV, MeV, GeV) – both for the energy they can give a particle using an accelerator, and for the rest mass of those particles. The electron rest mass is 511 keV/c^2, which works out to 9.1×10^{-31} kilogrammes. The proton's is 938 MeV/c^2, about 1.67×10^{-27} kg. Because energy and mass are almost synonymous, particle physicists usually drop the factor of c^2 and speak of particle masses simply in MeV or GeV.

2
Boson power

Two particle groups, the quarks and leptons, come together to form matter. A third group, the bosons, binds them and rules their lives in other ways.

The four fundamental forces

We now know that the great diversity of our universe stems from a few subatomic building blocks. Just as remarkable is the discovery that these particles interact in a few basic ways. There are four fundamental forces: gravity, electromagnetism and the weak and strong nuclear forces. Particle physics has so far failed to encompass the force of gravity – which in any case is profoundly weak in the kind of particle collisions we can create – but it has transformed our view of the other three. Through the lens of quantum physics, the forces that shape our world and guide the dance of every subatomic particle are themselves the effect of particles.

Electromagnetism

We have all felt the forces of electricity and magnetism. If you try to bring the north poles of two magnets together, you will find that there is a repulsive force. Rub a balloon on your sweater and hold it up to the ceiling. The balloon will stay there, held by an attractive electrical force.

In 1785 the French physicist Charles Coulomb worked out that the force between electrically charged objects follows an inverse square law (as Newton had shown for gravity a century earlier). The force, F, is proportional to the product of the two charges, p and q, divided by r^2, the square of the distance between them.

In the nineteenth century the English scientist Michael Faraday invented the idea of fields of force. Imagine an electrically charged object hanging in the centre of a room and suppose you have another object with the same type of charge, and that you can measure the force between the two objects at any point in the room. The direction of the force will always be along the straight line joining the two objects. So you could map the direction of the force as lines radiating from the central object.

The field around a magnet becomes visible if we use iron filings and allow them to line up in the direction of the magnetic forces. Physicists can also represent these fields mathematically, with equations that give the strength and direction of a force at all points in space and time.

Electric and magnetic forces are inextricably linked. For example, moving electric charges produce magnetic effects, while moving magnets can induce electric currents. In the mid-nineteenth century the Scottish physicist James Clerk Maxwell distilled all the experimental results and laws of electricity and magnetism into the single theory of electromagnetism: four equations describing the behaviour of an overall electromagnetic field.

Maxwell's laws are not enough, however, to explain in detail the reactions between atoms and molecules, or the structure of an individual atom and the way that it absorbs or emits light. To work at the subatomic level, we must allow for two new factors. Electrons in atoms move at high speeds, so they are subject to Albert Einstein's special theory of relativity. And this is a realm ruled by quantum mechanics, which gives many counterintuitive properties to particles. For example, particles act like waves and they can be in more than one state or position at the same time. Because electrons behave as waves, they are restricted to certain fixed wavelengths and therefore fixed energies when orbiting an atom.

The theory of electromagnetism that incorporates both special relativity and quantum mechanics is quantum electrodynamics, or QED. It was developed in the 1920s, by the British physicist Paul Dirac and others, and perfected in the 1940s. QED is a quantum field theory. The electromagnetic field is quantized, meaning that it can be treated as a collection of particles, or quanta. All interactions involve the emission and absorption of these quanta. Two electrons, for example, interact through the emission and absorption of a photon. It's rather like a game of quantum catch where the photon is the ball.

However, there is an important difference. The quantum ball's existence can only be temporary, as otherwise it would flout the basic principle of conservation of energy. It lives only briefly in what physicists call a virtual state. The full process must include both the emission and the absorption of this virtual particle.

So in QED, two protons, for example, repel one another by exchanging virtual photons. But virtual photons themselves can exchange more virtual photons, and to work out the strength of the force you have to add up every possible pattern of particle exchange. These are visualized in simple diagrams devised by US physicist Richard Feynman (see Figure 2.1).

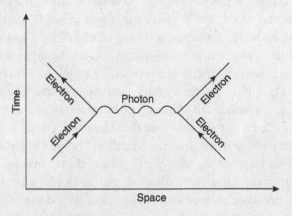

FIGURE 2.1 A simple Feynman diagram showing two electrons repelling each other by exchanging a virtual photon

The strong force

On the subatomic scale, two more forces come into play. There has to be a binding force between neutrons and protons, or else the powerful electric repulsion between protons would blow atomic nuclei apart. This force has no measurable influence outside the nucleus, so, unlike gravity or electromagnetism, it must be limited to a very short range – only about 10^{-15} metres. The nuclear binding force is about a hundred times as strong as the electric force between protons.

It is also much more complicated. Accelerator experiments show that this is not just a simple inverse square law, but is repulsive at the shortest distances, then attractive, then fades away rapidly as distance increases further. It also depends on how particle spins are aligned. Perhaps this should be no surprise. Protons and neutrons are composed of quarks, and it is a force between quarks that indirectly gives rise to the nuclear force between protons and neutrons.

Quarks have a property akin to electric charge, called their colour charge. It has nothing to do with colour in the everyday sense of the word, but it comes in three varieties, named red, green and blue. Just as mixing three colours of light gives white, so the three colour charges of quarks can add up to give no colour. This is what happens in a proton, for example, which always contains one blue-charged quark, one red and one green.

The strong force can still hold protons and neutrons together in the nucleus, in the same way that electromagnetic forces can hold electrically neutral atoms together. The positively charged nucleus of an atom is only partly screened by its own electron cloud and, similarly, the coloured quarks inside a proton can feel the presence of quarks in the proton next door.

Because this whole theory is a quantum field theory like QED, but involves so-called colour charges, it is known as quantum chromodynamics, or QCD. In QCD the particles that carry the strong force are called gluons, because they glue particles together.

There is one important difference between gluons and photons. Photons are not electrically charged, and so do not interact with each other directly through the electromagnetic force (only indirectly, via charged particles). But gluons carry their own colour charges, so they can interact with each other via the strong force. This limits the range of the strong force and makes it behave in a peculiar way.

When two electrically charged particles are pulled apart, the force between them decreases; but when you move two quarks apart, the gluons pull on each other and the force actually gets stronger. As far as we know, a lone quark can never escape from inside a proton, but must always exist in a colourless combination with other quarks. During a collision, energy materializes not as single quarks but as quark–antiquark pairs – mesons.

Colour belongs only to quarks. Leptons, such as the electron, have no colour and so they do not feel the strong force at all.

Forces that matter

- Residual effects of the electromagnetic force bind electrically neutral atoms together in molecules.
- The electromagnetic force holds the atom together, binding a cloud of electrons around a central nucleus.
- Residual effects of the strong force bind protons and neutrons together in nuclei.
- The strong force holds quarks together within protons and neutrons.

The weak force

A neutron that is free from the confines of an atomic nucleus does not live for ever. It will soon spit out an electron and a particle called an antineutrino, and become a proton. Physicists can describe this decay of the neutron in terms of a fundamental force, called the weak force, which is about 10,000 times weaker than the strong force. Unlike gravity and magnetism, which a human being can feel, it does not seem like a force in the everyday sense of the word. Rather than holding matter together – as the other forces do on different scales – the weak force allows the basic quarks and leptons to change from one type to another. So a top quark can become a bottom quark, or a muon can become an electron. The weak force is the only force that can do this. It also causes radioactive beta decay.

The quantum field theory of the weak force needs three carrier particles. The W+ and W- are electrically charged, and come into play when the weak force changes the charge of a particle, as when a neutron decays into a proton. The $Z°$ is uncharged, and mediates weak interactions in which there is no charge change.

All three carriers of the weak force are heavy, with masses roughly a hundred times that of a proton or a neutron. So how can a neutron emit a virtual W or Z particle that is much heavier than itself? The answer lies in quantum uncertainty. Quantum physics shows us that there is always some uncertainty in the mass (strictly, mass-energy) of a particle, or even of a point in empty space. Over a long period of time, this uncertainty is small; but over a short time, the uncertainty is large. A neutron can create a virtual W or Z particle out of nothing at all, provided that the particle is then absorbed by the neutron or another particle in a short enough time.

Photons, the field quanta of the electromagnetic force, have zero rest mass, so virtual photons can exist indefinitely and travel any distance, meaning that the electromagnetic force has an infinite range. But W and Z particles are heavy, and so cannot travel far from their parents. Like the strong force but for a different reason, the weak force therefore has a very short range.

You cannot detect a virtual particle. But in 1984 researchers at CERN collided beams of subatomic particles in the Super Proton Synchrotron to create real, rather than virtual, W and Z particles. These experiments confirmed a remarkable theoretical prediction. In the 1960s the physicists Sheldon Glashow, Abdus Salam and Steven Weinberg developed a quantum field theory for the weak force, which showed that at high enough energies electromagnetism and the weak force are parts of a single, unified 'electroweak' force. So when the universe was very young and space was filled with energetic radiation at a temperature of billions of degrees, W and Z particles could be made as easily as photons.

The success of unifying two forces in one theory has encouraged physicists to search for a single 'theory of everything' describing the electroweak, strong and gravitational forces as facets of a single underlying force (see Chapter 9). Such a theory might reveal the nature of time, space and mass, and would provide insights into the first moments of the universe, when all four forces were indistinguishable.

While that remains an elusive dream, QED, describing the electroweak force, and QCD, describing the strong force, are already tremendously powerful. The underlying symmetry of this theory (see Figure 2.2) points to there being six quarks and six leptons, as observed. It also predicted the existence of the W and Z particles, as well as the odd boson out, the mysterious Higgs.

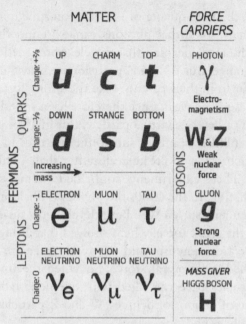

- Every **quark** and **lepton** has an antiparticle twin with the opposite electric charge.

- **Quarks** clump together as composite particles made either of a quark and antiquark (**mesons**) or of three quarks or three antiquarks (**baryons** such as protons and neutrons).

FIGURE 2.2 The panoply of particles within the standard model of particle physics explains the workings of all visible matter and three of the four fundamental forces. Quarks and gluons feel the strong force because they have colour charge, but leptons and the other bosons do not.

Bosons and fermions

Particles must obey the laws of quantum mechanics. For example, they have wave-like properties – moving as a quantum wave functions, spread out in space, with wavelengths that get shorter as their energies increase.

Many properties of particles are restricted to fixed, quantized values. These properties are assigned quantum numbers. One is spin, which can come only in integer or half-integer multiples of the basic quantum unit. Spin is similar in some ways to rotation in the classical world – it signifies angular momentum – but it also has profoundly quantum consequences.

A particle with half-integer spin, such as an electron, a proton or a quark (all with spin ½), has a type of asymmetry in its wave function that makes it antisocial. Known collectively as fermions, these particles cannot share a quantum state. This is behind the complexity of matter. It is the reason why putting six protons, six neutrons and six electrons together makes an interesting, structured atom of carbon-12.

A particle with integer spin, such as a photon (spin 1), is known as a boson (named after Indian physicist Satyendra Nath Bose). Bosons have symmetrical wave functions, and can happily sit in the same state as one another. They are no good for building complex matter, but instead play another role as carriers of force.

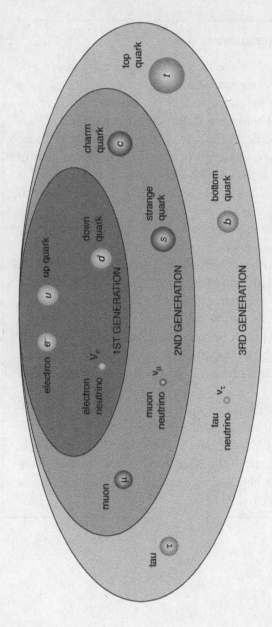

FIGURE 2.3 The particles that make up the standard model are grouped into three families, or generations, according to their flavour and mass.

Why symmetry rules the universe

An outsider might see physics as a morass of equations and particles, but those in the business see it as an elegant description of the universe, always aiming to be as simple as possible. To gain some idea of why, you need to understand what physicists mean by the word 'symmetry', and the stunning insight of a largely forgotten mathematician, Amalie 'Emmy' Noether (see Figure 2.4), which laid the groundwork for nearly every major fundamental discovery since.

FIGURE 2.4 The German mathematician Emmy Noether (1882–1935), in a photograph taken before she entered the University of Göttingen

The mathematician Hermann Weyl defined the term 'symmetry' like this: 'A thing is symmetrical if there is something you can do to it so that after you have finished doing it, it looks the same as before.' A circle, for instance, can be rotated by any angle and looks the same.

The idea that symmetry lies at the heart of physical laws is an old one. Aristotle and his contemporaries argued that the stars were pasted on celestial spheres, and that the globes moved in circular orbits. They were wrong, of course; but when Newton explained the elliptical paths of planets with his law of gravitation, he introduced a new symmetry – the symmetry of the invisible hand of gravity, which acts equally in all directions from a massive body such as the Sun. General relativity, Einstein's much refined theory of gravity, was founded on a symmetry known as the equivalence principle: that there is no discernible difference between a body experiencing acceleration due to gravity and one experiencing an equivalent acceleration from a different source, such as the thrust of a rocket or the spin of a centrifuge.

Einstein's work spurred a great deal of interest in the role of symmetries in physical laws and, recognizing Noether as an expert, the mathematicians David Hilbert and Felix Klein invited her to Göttingen in 1915. Almost immediately, Noether developed her eponymous theorem. Put simply, it says that symmetries give rise to conservation laws.

Noether's theorem

Conservation laws are the bread and butter of physics. They are mathematical shortcuts that allow us to compute physical quantities once and then never again. Whatever you start with, that's what you'll end up with. Most of the great laws of physics include some statement of conservation, implicitly or explicitly. Newton's first law of motion crudely states that

SYMMETRY:
TRANSLATION IN TIME
The basic laws of physics do not vary over time.

CONSEQUENCE:
Energy is conserved
However many times a pendulum swings, with no friction it will always reach the same height.

SYMMETRY:
TRANSLATION IN SPACE
The laws of physics don't change when you move from one place to another.

CONSEQUENCE:
Momentum is conserved
A rocket flying through empty space continues to move at the same velocity.

SYMMETRY:
ROTATION IN SPACE
Forces such as gravity emanate equally in all directions.

CONSEQUENCE:
Angular momentum is conserved
Comets speed up nearer the sun. The area between their path and the sun is always the same in a set time.

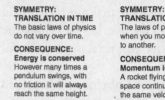
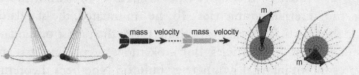

FIGURE 2.5 Symmetries exist everywhere in nature: Emmy Noether's theorem of 1915 provides a way to translate them into laws useful for calculations.

'objects in motion stay in motion, and objects at rest stay at rest'. That is conservation of momentum. And Noether's theorem (see Figure 2.5) tells us why it holds true.

Consider a hockey puck placed on a very smooth, very large frozen lake. Wherever the puck slides, the lake is the same. Apply Noether's theorem to that particular spatial symmetry, and it tells you that momentum is conserved. (The conservation law only holds as far as the symmetry does; a hole in the ice will disturb the symmetry, causing the puck to sink to the bottom of the lake.)

It is not always obvious what is conserved and what is not. Before Noether, energy was simply assumed to be conserved, an assumption so basic that it became known in the nineteenth century as the first law of thermodynamics. But if you do the mathematics associated with Noether's theorem, it becomes plain that energy is conserved because of an even more basic symmetry: specifically, that the laws of physics are not changing with time. If they did, energy would not be conserved.

Noether's theorem is a prescription for making progress in physics. If you identify a symmetry in the world's workings, the associated conservation law will allow you to start meaningful calculation.

Symmetries in space and time might be obvious to the naked eye, yet Noether's theorem's true strength comes from more obscure 'internal symmetries'. To the uninitiated, the standard model of particle physics is nothing more than a list of fundamental forces and particles. But look more deeply and it is an expression of internal symmetries, built on Noether's theorem.

Take electromagnetism. James Clerk Maxwell is generally credited for writing down a theory that unified electricity and magnetism into one working model in the 1860s. One of its assumptions is that electric charge is neither created nor destroyed – and Noether's theorem shows that charge conservation arises from symmetry (see Figure 2.6).

In a spin

Fundamental particles have a property called spin, and just as position doesn't matter on a frozen lake, the spin's phase does not change physical calculations. Turn every electron in the universe an extra degree, and neither energy nor anything else changes. According to Noether's mathematics, what pops out of this internal symmetry is charge conservation.

Weyl took this idea of phase symmetry a step further and supposed that every electron might be twisted by a different amount without changing any measurable quantity. Almost by magic, all four of Maxwell's equations emerged. As the standard model has developed, the symmetries of interest have become more subtle, but Noether's theorem has been the gift that keeps on giving.

It is hard to conceive that electrons, the particles that run through wires to power electronics, and neutrinos, which fly

FIGURE 2.6 For physicists, the appeal of symmetry goes beyond the
purely aesthetic.

through us by the trillions every second without leaving a mark,
are in some sense the same particle. Neutrinos primarily interact through the weak force, which controls nuclear fusion in
the Sun. But the weak force is indifferent to whether a particle
is an electron or a neutrino: switch them round and weak interactions will be the same. This symmetry produces conservation
of a quantity called weak isospin that, like electric charge, can
be used to label particles and predict how they will behave.

In the 1960s researchers found that electromagnetism and the
weak force could in fact be generated by a single underlying symmetry, in what became known as the electroweak theory, a key-
stone of the standard model (see Figure 2.7). When the symmetry
is unbroken, at very high energies electrons and neutrinos behave
identically. In today's cool universe the symmetry is broken, which
changes the behaviour of electrons and neutrinos and produces a
new particle – what we now know as the Higgs boson.

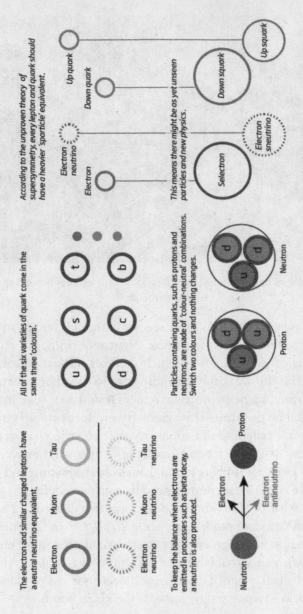

FIGURE 2.7 The workings of the standard model of particle physics – and perhaps theories beyond it – are determined by some subtle symmetries.

The other pillar of the standard model is the strong interaction, which holds individual protons and neutrons together. The quarks that make up these particles are labelled with one of three 'colours': red, green and blue. Shift all the colours by one, and all strong interactions will remain exactly the same. This symmetry leads to the conservation of colour, a principle that constrains what particles can exist and what decay processes are possible.

New theories in particle physics are built on informed guesses about the fundamental symmetry of the universe. The Holy Grail is unification: the drive to develop theories that can describe everything in just a few equations. For decades, theorists have thought that this final symmetry should include something called supersymmetry, which posits heavy boson partners for all fermions, and heavy fermions for all bosons (see Chapter 8). The desire to create and observe these superpartners was one of the main motivations for building the Large Hadron Collider, which, viewed in the right light, is just a big machine for seeing symmetry.

Gender asymmetry

Despite wide recognition of her brilliance, Emmy Noether was confounded by the prejudices of academic tradition. Born into a prominent mathematical family in 1882 – her father, Max, was a professor at the University of Erlangen in the north of Bavaria – she was at first forbidden from enrolling at the university because of her gender.

Even though Noether was eventually able to gain both an undergraduate degree and a PhD, still no university would hire her for their faculty. Over the next decade, she became one of the world's experts in the mathematics of symmetry, but without appointment, pay or formal

---

title. Even when she was invited to work at Göttingen, the offer came without remuneration. For seven years she was allowed to serve as a guest lecturer, until receiving an honorary 'extraordinary' professorship in 1922.

Herman Weyl, also at Göttingen in the 1920s, by contrast quickly achieved a prominent professorship, despite being Noether's junior. 'I was ashamed to occupy such a preferred position beside her whom I knew to be my superior as a mathematician in many respects,' he later remarked.

Noether left Germany to escape the Nazis in 1933 and came to Bryn Mawr College in Pennsylvania, dying of complications from cancer surgery two years later. As Einstein wrote after her death, 'Fräulein Noether was the most significant creative mathematical genius thus far produced since the higher education of women began.' Others might suggest that the second part of that sentence is superfluous. Mathematicians revere her, yet, despite the fact that she laid the groundwork for much of modern physics, physicists tend to gloss over her contributions.

Why do we need the Higgs?

The standard model (see Figure 2.2) is our most successful theory of reality. It describes how particles of matter – fermions – feel forces and interact through the exchange of other particles – bosons. But for decades, one crucial component was notable by its absence: the Higgs boson. This particle is thought to play two essential roles: giving other particles mass and explaining why nature's forces take the form they do.

If you break matter up into smaller and smaller pieces, whatever you start with you eventually end up with a bunch of particles like electrons and a bunch of quarks: the lighter up quarks and down quarks that make up protons and neutrons, and their short-lived cousins, the strange, charm, bottom and top quarks. The electron belongs to a different family of six particles, the leptons, together with its two heavier cousins, the muon and the tau, and three near-massless neutrinos that partner each of these. These 12 matter particles, collectively known as fermions, have an anti-particle partner that is identical, except that it has the opposite charge. And that's it. As far as we know, matter cannot be broken down any further.

This neat pattern fits the experimental facts, but hides a perplexing problem. All matter particles have a property called mass – a resistance to changing speed and direction. Their masses range over more than 11 orders of magnitude, from the lightweight electron neutrino to the humongous top quark (see Figure 2.8). Where do these masses come from – and why are they so different?

This is far more than a matter of particle physics bookkeeping. Without something to confer mass, there would be no matter as we know it. Massless quarks and electrons would constantly move at the speed of light and it would be impossible to bind them into protons, neutrons and atoms.

According to the standard model, fermions interact through forces transmitted by other particles, known as bosons. Photons latch on to electric charge and carry the electromagnetic force, while gluons feel colour charges to produce the strong nuclear force.

As for the third of the standard model's forces, the weak nuclear force ... well, it is weak, but without it the radioactive decay that powers the Sun and other stars would not occur. Its weakness comes about because its carriers, the W and Z

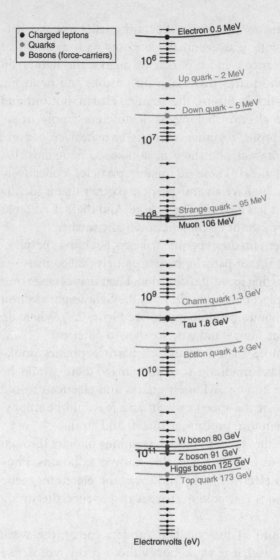

FIGURE 2.8 Masses of elementary particles. The masses of the three neutrinos (electron neutrino, muon neutrino and tau neutrino) are very small, much less than 1 eV, and not shown in this diagram. Photons and gluons have no mass.

bosons, have very large masses – almost 100 times the mass of the proton. Creating such particles takes a lot of energy. Under normal conditions, matter particles prefer to interact by swapping massless photons, if they can.

At very high energies – in the first split-second of the universe, for example, or in collisions in powerful particle accelerators – this difference melts away. The W and Z particles become massless like the photon, and the electromagnetic and weak forces, so hugely different in our everyday experience, become unified into the electroweak force.

The process by which the electroweak force splits into the electromagnetic and weak forces is known as electroweak symmetry breaking, and must have happened some time in the universe's early moments. Whatever caused it is connected to the mystery of mass. After all, it is the mechanism by which the W and Z bosons acquired mass.

Broken symmetries are not restricted to exotic forces. An everyday example is when a liquid cools into a solid crystal. Here, a broadly symmetrical state (everything looks the same in all directions in a liquid) is replaced by a less symmetrical state, in which things look distinctly different along different axes.

In the 1960s particle theorists began to wonder whether tools developed to describe this symmetry breaking could be applied to the cooling cosmos. This was no easy task. Molecular interactions in a solid or liquid can be defined by reference to a fixed set of co-ordinates, but thanks to the space-time warpings of Einstein's general relativity, there is no such standard frame of reference for the universe.

In 1964 the Belgian theorists Robert Brout and François Englert devised the equations of a quantum field that could break symmetry in a way consistent with relativity, and give mass to particles. The British physicist Peter Higgs made the same proposal, and pointed out that ripples in this field would take the form of a new particle. Later that same year, Gerald Guralnik, Carl Hagen and Tom Kibble combined these ideas into a more realistic theory that was a precursor to the standard model.

The point about the Higgs field is that even the lowest-energy state of space is not empty. Particles travelling through space interact with the field to different degrees, and this creates a 'sticky' quality to their movement: mass. The W and Z bosons acquire their mass by one kind of interaction with this field, while fermions do so by another kind of interaction. Because the Higgs field has no net electric or colour charge, photons and gluons do not interact with it at all, and so remain massless.

This was a neat trick. To find out if it worked, we needed to expose the field by giving it a good hard kick to make it wobble. Those wobbles could then be observed as Higgs bosons. Theoretical and experimental developments gave us a good idea of the energy required: the Higgs boson's mass had to be between about 100 and 400 gigaelectronvolts. We would need a truly huge particle accelerator.

What does a quark–gluon fireball sound like?

The hottest material ever created in the laboratory makes an eerie drone. A similar sound may have pervaded the universe just after the Big Bang, when space was a seething cauldron of matter.

The lab-made material was created at the Relativistic Heavy Ion Collider in Upton, New York State. This accelerator slams gold ions together, breaking the atoms and their constituent protons and neutrons into even smaller bits called quarks and gluons. The resulting

fireball – called a quark–gluon plasma – has a temperature of trillions of degrees and mimics conditions when
the universe was a millionth of a second old. As the fireball created in this 'little bang' cools, the individual quarks
and gluons combine into a zoo of larger particles.

Physicist Ágnes Mócsy of the Pratt Institute in Brook-
lyn, New York, and colleagues have calculated what this
fireball of quarks and gluons would sound like to an
observer embedded within it. By analysing measurements
made using roughly 3 million collisions, the team determined the general lumpiness of the fireballs – how closely
spaced their particles were.

Fluctuations in density correspond to sound waves. So
the researchers studied how the distribution of particles
evolved in time to see how the sound changed. They then
had to multiply the wavelengths of the sound by roughly
10 billion billion to be audible to the human ear. In the
resulting soundtrack, lower tones become more and more
prominent as the fireball expands and the speed of sound
changes due to the resulting drop in the fireball's density.
About halfway through, a wiggle in the tone signals the
point at which quarks and gluons recombine to form particles from protons to pions.

To hear the soundtrack, visit www.newscientist.com/
article/dn18998-what-does-the-hottest-matter-ever-
made-sound-like/

3
The Higgs maker

The Large Hadron Collider (LHC), near Geneva in Switzerland, is the biggest ever physics experiment. By studying conditions similar to those in the early moments of the Big Bang, it could solve some of science's deepest mysteries and take us into new realms of particle physics. It was switched on in September 2008 and four years later it revealed the last piece of the standard model, the Higgs boson.

Introducing the LHC

Straddling the border between France and Switzerland (see Figure 3.1), the LHC is designed to answer some of the most profound questions about the universe. What is the origin of mass? Why are we made of matter and not antimatter? What is dark matter made of? The LHC could also provide clues about conditions in the very early universe, when the four forces of nature were rolled into one giant superforce.

To find answers to these questions, the LHC sends protons travelling at 99.9999991 per cent of the speed of light around a circular tunnel. It then smashes them together at four points on the ring, each of which is surrounded by a huge detector. The collision energy produced is 14 teraelectronvolts (TeV), seven times greater than its predecessor the Tevatron, at Fermilab in Batavia, Illinois.

In everyday terms, this energy is not so great. A flying mosquito has about 1 TeV of kinetic energy. What makes the LHC special is that this energy is concentrated in a region a thousand billion times smaller than a speck of dust. So great is the concentration of energy at the LHC that it recreates conditions similar

FIGURE 3.1 The Large Hadron Collider straddles the Swiss–French border.

to those 10^{-25} seconds after the first moment of the Big Bang, soon after the particles and forces that shape our universe came into being. With so much energy available, the LHC should be able to create new massive particles for the first time in the lab.

It has already succeeded with the Higgs boson, thought to give all other fundamental particles mass. But physicists are still hoping for more, including supersymmetric particles that point the way to new physics beyond the established standard model. The lightest supersymmetric particle is also a promising candidate for dark matter, the invisible entity thought to amount to 95 per cent of the universe's mass. Some theorists speculate about more outlandish discoveries for the LHC, including the production of extra dimensions, mini black holes, new forces, and particles smaller than quarks and electrons. Although the LHC has seen none of these yet, there is still a chance that they might appear over the next few years. It is also still searching for subtle differences between matter and antimatter, and studying an exotic state of matter called quark–gluon plasma, which could reveal how the quarks and gluons from the Big Bang fireball condensed into the protons and neutrons we see today.

Could the LHC or another particle accelerator make a black hole that swallows the Earth?

In 2008 campaigners in the US attempted to delay the start-up of the Large Hadron Collider, claiming that it might spawn mini black holes that would destroy the entire Earth. Could such a machine really generate black holes?

The answer is no. If black holes were created at the LHC, they would evaporate in 10^{-26} seconds, via a process first described by British physicist Stephen Hawking. Even if Hawking is wrong and black holes do not evaporate, we have experimental reasons for feeling safe. Cosmic rays from outer space have far greater energies than the LHC can produce and have been colliding with the solar system's planets for billions of years without problems. What's more, there are far more cosmic-ray collisions than LHC collisions. No black holes have swallowed Jupiter or Saturn or the Moon.

The making of the extreme machine

At 27 kilometres long, the LHC is the largest machine in the world. It accelerates beams of protons to within a whisker of the speed of light and smashes them head on, 600 million times a second. Perhaps the most distinctive feature of the LHC, though, is its temperature. At just 1.9 kelvin, the LHC is the coldest thing of its size in the universe ... unless an alien civilization has built a colder one.

Were it not for this searing cold, the LHC might have suffered the same fate as the Superconducting Super Collider (SSC), which was on its way to becoming the most powerful accelerator in the world in the early 1990s. The SSC was designed to smash protons into each other at energies of 40 TeV. This requires intense magnetic fields generated by superconducting magnets to bend and focus the beam (see Chapter 1). The SSC team opted not to push the technology; they chose superconducting magnets of a type already being used in other accelerators, cooled by liquid helium to 4.5 K. These magnets are strong but not as strong as the cooler ones developed for the LHC – and that proved fatal. Weaker magnets cannot bend such energetic particles into a tight circle, so the SSC's tunnel had to be 87 kilometres in circumference. The cost of building a machine this big doomed the SSC, and the US government cancelled it in 1993. The SSC's partially completed tunnel, near Waxahachie, Texas, now lies derelict.

Keen to avoid a similar debacle, CERN, the European particle physics lab near Geneva, Switzerland, took the decision to cram the LHC into an existing tunnel 100 metres underground, built in the 1980s for the Large Electron–Positron Collider (LEP). It is commonly described as a ring but is actually more like an octagon with rounded corners. The designers chose a less ambitious peak energy of 14 TeV – still thought to be enough to reveal plenty of new physics – and next-generation superconducting coils made of niobium–titanium. Cooling these coils down to 1.9 K meant that they could carry much more current and generate the extra-strong magnetic fields the machine needed. But it came at a price. At that temperature, liquid helium becomes superfluid. With zero viscosity it can slip through microscopic cracks. So the thousands of welds in the plumbing had to be as good as those in a nuclear plant.

Was the LHC taken down by a weasel?

No – it was actually a beech marten. On the morning of Friday, 26 April 2016, the creature chewed on a 66-kilovolt electrical transformer, causing a CERN-wide power cut. Originally called a weasel in the logbook, it was later more accurately identified as a beech marten, a larger relative of the weasel. It took a week to bring the collider back online. The marten was put out of action permanently.

The incident is reminiscent of a story widely reported in 2009 that a power cut at the LHC was caused when a bird dropped a piece of baguette on a substation – although this has never been confirmed.

ATLAS and CMS detectors

The beam line, which the particles fly along, is only part of this great machine. To find out what happens when particles collide, four detectors are installed around the ring (see 'Detector story' below). ATLAS is the largest of these, a 7,000-tonne detector that tracks the particles flying out from the collisions. Among the most important of these are muons, and the way to measure their momentum is to bend their paths in a magnetic field. Because of the LHC's power, these muons are more energetic and faster moving than anything seen in previous colliders, so the magnetic fields have to be very strong. ATLAS ended up with the world's largest superconducting magnet, by volume.

The cavern created to house this behemoth had to be 35 metres high. It's so big that it is rising through the denser rock around it rather like a bubble in water, albeit extremely slowly. It moves about 0.2 millimetres upwards per year, and the floor had to be cast 5 metres thick to ensure that it will not warp as it rises.

To cross-check the findings from ATLAS, a second catch-all experiment called the Compact Muon Solenoid (CMS) hunts for the same particles using different technology (see Figure 3.2). It has thrown up its own share of challenges. As protons collide inside the detectors, the aftermath can slightly disrupt the path of other protons as they race around the ring. To minimize this effect, the detectors have to be as far away from each other as possible. So the CMS is sited diametrically opposite ATLAS, which puts it at the base of the Jura Mountains. This spelled trouble. First, the engineers had to dig two 60-metre-deep shafts, one for elevators and one to lower the detector. But they found that the area consisted of loose

FIGURE 3.2 The Compact Muon Solenoid (CMS) experiment is one of two large general-purpose particle physics detectors built on the proton–proton Large Hadron Collider (LHC) at CERN.

gravelly sediment with water flowing through it – hopeless for digging. So they froze the ground.

They drilled out a series of pipes and circulated brine through them at −5 °C; then, for a month, they filled the pipes with liquid nitrogen at -196 °C. This created a 3-metre-thick retaining wall of ice that kept the groundwater at bay, while the workers dug the dry earth within.

Meanwhile, the CMS engineers were working on the world's most powerful superconducting magnet, twice as strong as the one in ATLAS. The 10,000-tonne magnet is based on superconducting coils that must withstand an outward force of 60 atmospheres, generated by the magnet's 4-tesla field (about 100,000 times stronger than the Earth's field). No one company, or country, could do the job, so the magnet's coils were shunted around Europe on an eight-year journey that started in Finland and took in Switzerland, France and Italy en route to CERN.

It was crucial that the entire LHC and its detectors worked from the word go, as repairing them once the system was up and running would be far from trivial. A repair would mean letting the LHC warm back up to room temperature, which takes about five weeks. Afterwards, its 40,000 tonnes of magnets would need to be cooled back to 1.9 K, a process that takes another five weeks and requires nearly 10,000 tonnes of liquid nitrogen and 130 tonnes of superfluid helium. Not surprisingly, quality control was tight.

If the physical LHC itself is revolutionary, so too are the computing systems that collect, scrutinize and store the resulting data. Each high-energy collision sprays out a plethora of particles that need to be tracked and identified. It is impossible to record them all, so smart software makes the selection, discarding more than 99.99 per cent of all collisions. Even then, the LHC's four experiments will generate 15 million gigabytes of data a year. To cope with this onslaught, engineers had to create one of the most sophisticated data handling and analysis systems ever to exist, the Worldwide LHC Computing Grid, which links more than 140 computing centres in 34 countries.

Detector story

Each of the LHC's four large detectors (ATLAS, CMS, LHCb and ALICE) is built up of concentric cylinders. Cylinders closest to the collision point are made of a semiconductor. When an electrically charged particle passes through this material, it liberates loosely bound atomic electrons, creating a pattern of electrical currents that shows the particle's path. Magnets surrounding the detector cause the paths of the charged particles to curve, and the degree of curvature reveals the particle's momentum.

The next cylinder out consists principally of detectors based on liquid argon (in ATLAS) or lead tungstate (CMS). Collisions with the densely packed atoms in these detectors stop most particles in their tracks, and generate photons that are used to measure each particle's energy, and so identify it.

Muons are not stopped by these detectors, but are identified and measured by the next cylinder of dedicated detectors. Weakly interacting neutrinos are not measured at all; their presence is instead deduced by totting up the momentum of all the other particles produced in a collision and seeing whether anything is left unaccounted for.

The products of many simultaneous proton–proton collisions fly away at close to the speed of light, and collisions that need a closer look need to be picked out as quickly as possible because, within 25 nanoseconds, another bunch of protons will collide at the detector's heart.

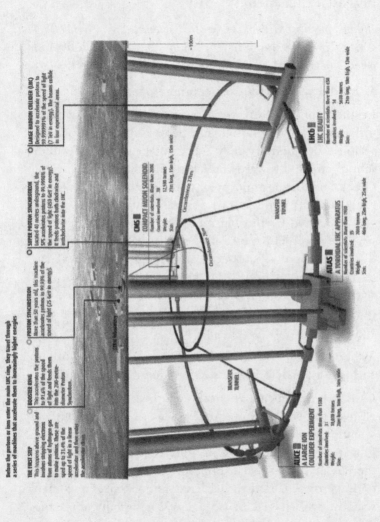

FIGURE 3.3 The Large Hadron Collider accelerates beams of protons (or lead ions) in opposite directions, and collides them head on at four locations where huge detectors analyse the debris.

Interview: The woman in charge of the world's biggest experiment

Italian physicist Fabiola Gianotti became director-general of CERN in 2016. New Scientist *interviewed her in 2009, just before she took over as head of the ATLAS detector.*

Why did you decide to become a particle physicist?

I came to physics from very far away. When I was a young girl, I loved art and music. I had been studying piano quite seriously at a conservatory and had taken courses in high school targeted towards literature, languages like ancient Greek and Latin, philosophy and history of art. I loved these subjects but I was also a very curious little girl. I was fascinated by the big questions. Why are things the way they are? This possibility of answering fundamental questions has always attracted me – my mind, my spirit, everything. So when I had to choose what I wanted to do with my life, I thought that physics could answer these big questions in a more concrete way than philosophy. I was right – in that I'm very happy now.

What are your thoughts on becoming the first woman to head a particle physics experiment at the LHC?

CERN is such a rich environment: there are people from all over the world, young students work with established scientists and Nobel prize-winners. So geographical origin, age and gender make no difference here. I don't feel there is anything special about a woman leading a big scientific project. On the other hand, I hope that as a woman scientist who has achieved a level of visibility in a big experiment like ATLAS, I can be an encouragement to young women who are thinking of a scientific career.

Give us a sense of the size and scope of ATLAS.

The ATLAS collaboration consists of almost 3,000 physicists from 169 institutions. ATLAS is the biggest detector ever built at a particle collider and its spectacular size strikes people immediately when they visit the underground cavern it is housed in – it's as big as a five-storey building. This size is combined with an enormous complexity. There are 100 million independent electronic signals that we need to record in order to reconstruct the hundreds of particles produced in every proton–proton collision. The trajectories of the particles must be reconstructed with micrometre precision. This amazing combination of size, complexity and precision has made the technology very challenging. ATLAS and indeed the other detectors at the LHC are instruments without precedent.

What are the key goals of ATLAS?

ATLAS will sift through particles created by extremely high-energy proton–proton collisions. We are starting on a fantastic scientific journey. We believe that at this energy scale, new physics should manifest itself, physics beyond the so-called standard model [which explains all known particles and the forces that act upon them]. We expect to find answers to some fundamental questions and mysteries, many of which have been with us for decades. For instance, what is the origin of mass? It's a question related to the existence of the Higgs boson. Are there other forces of nature, in addition to the four forces we already know of? Are there additional dimensions of space? What is the composition of the universe's dark matter?

What would you personally like to see ATLAS discover first?

Dark matter. I would be very, very happy if we discover the particle that explains 20 per cent of the universe's composition. Accelerators like the LHC allow us to study the infinitely small – the basic constituents of matter – and this can tell us about the structure and evolution of the universe, stressing the link between the infinitely small and the infinitely big.

Have you thought about what would happen if no new physics is discovered at the LHC?

It is a good question, but it's difficult to answer. Based on what we have learned from experimental and theoretical work over the last few decades, there must be something new at the energy scales that the LHC will offer. Perhaps there will be just one Higgs boson, or a new mechanism playing the same role, but we expect more. We know that the standard model is not a complete theory of elementary particles, because it cannot answer all our questions. We expect it to start to break down at the energy scale of the LHC. There must be new physics there. Perhaps they won't be the answers that we have in mind, but there must be answers. Nature could well present us with surprises and this will be one of the most exciting possibilities. After all, research is about looking for something that we don't know a priori.

What has it been like working on ATLAS?

Building the LHC and experiments like ATLAS is an unprecedented scientific, technological and human adventure. What makes my life as a scientist at CERN so special is the combination of three elements. One is the exciting physics goals. Then, to address these questions, we had to develop high-tech instruments which are at the cutting edge of technology in various sectors, from electronics to cryogenics, and which have spin-off benefits to society. Thirdly, these projects have been carried out in an international environment, with physicists, engineers and technicians from all over the world, bringing nations together through science and breaking political barriers. In our project, we have people from countries that are historically not the best of friends.

The LHC and ATLAS could reveal some deep truths about how the universe works. What do you feel when you think about it?

Feelings of excitement, of course, and the awareness of being close to something very important and great for humankind. Fundamental research is a duty and a need of human beings. The thirteenth-century Italian poet Dante said that we were not created to live as animals but to pursue virtue and knowledge. As human beings, the pursuit of fundamental research and knowledge is a need for us, which separates us from animals or vegetables. It is like the need for art. Research brings knowledge, and knowledge brings progress, always.

If we discover something fundamental at the LHC, it will be a bit like going to the heart of the universe. When you are getting closer to the fundamentals, to the basic questions of where the universe comes from and where it is going, there's a very special feeling.

The big discovery

In July 2012 the LHC team announced that they had finally found the Higgs boson – but they didn't see it directly. The Higgs is short-lived, decaying almost instantaneously into other particles. To infer its presence, you must measure these decay products and look for evidence that they came from a Higgs.

Should the Higgs boson really be called that?

In 2012 organizers of a physics meeting requested that the Higgs boson instead be referred to as either the BEH or scalar boson. The name change might seem esoteric, but it hints at a complex past and difficulties over whom to credit for the discovery of this particle.

To understand, rewind more than 50 years. As with most scientific advances, a single mind did not solve this puzzle. Work by Yoichiro Nambu of the University of Chicago in 1961 led to the idea that a mass-giving field interrupted an early universe until then filled only with massless particles.

In August 1964 Robert Brout and François Englert (the 'B' and 'E' in BEH) at the Free University in Brussels, Belgium, ironed out some kinks in the theory and detailed a mechanism. But it was Peter Higgs at the University of Edinburgh, UK, who first explicitly predicted the particle we now call the Higgs – in a paper published in October 1964.

This progression explains the BEH boson option. But others favour a more anonymous name: the scalar boson or, alternatively, an even complicated name – the BEH-HGK boson, which gives credit to credit Dick Hagen, Gerald Guralnik and Tom Kibble, who published a mass-giving mechanism in 1964.

> Others point to the ad-hoc way in which particles seem to be named. 'It's not like elements whose names are very carefully chosen by a committee,' says physics Nobel laureate Steven Weinberg at the University of Texas at Austin, who named the Z boson that carries the weak force.

Fortunately, the standard model predicts everything we need to know about the Higgs, apart from its precise mass. For every possible mass, we can predict the number of particles that the LHC should produce, and what they will decay into. For example, the Higgs should sometimes decay into pairs of high-energy photons. There are many possible sources of a photon pair, but if we concentrate on the likely looking ones and plot their combined momentum on a histogram, an unknown particle will make itself known as a 'bump' – an excess of events corresponding to a particular mass (see Figure 3.4). This is what both ATLAS and CMS saw at a mass of around 125 gigaelectronvolts, and announced to the world on 4 July 2012 as the Higgs boson.

That was not the only evidence. The Higgs boson was also expected to sometimes decay into two Z bosons, which each decay further to two leptons. Combining the momentum of these leptons produced a peak at the same mass as in the photon data. Decays to W bosons added their own strand of evidence. W particles decay into neutrinos, which are not detected, so there is no definite mass bump in this case. Instead, we just see more W decays than we would expect if the Higgs did not exist.

Indeed, the new particle decays to W and Z bosons at roughly the rate predicted for a Higgs by the standard model. So we know enough to call this a Higgs boson of some sort. But we still cannot be sure whether this is exactly the Higgs predicted by the standard model, or something even more interesting.

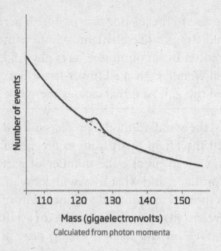

Mass (gigaelectronvolts)

Calculated from photon momenta

FIGURE 3.4 Measuring the momentum of photon pairs produced in collisions at the Large Hadron Collider revealed a suspicious bump – one line of evidence for a new particle.

Since the LHC powered up again in 2015, it has been probing several properties of the new particle. While we are pretty sure the new particle decays into force-carrying bosons as a standard-model Higgs should, we are less sure about decays into matter-making fermions. The upgraded LHC has been measuring these Higgs decays into bottom quarks, tau leptons and even muons.

The most intriguing questions surround the particle's mass. In the standard model, the interactions of the Higgs with itself and the particles around it seem to imply it should have a huge mass. The particle discovered by the LHC is far smaller. Order can be restored by 'fine-tuning' the standard model, tweaking things so that two large numbers almost, but not quite, cancel each other out, leaving the Higgs with a small mass. Many are unhappy with this fix, believing that it makes the theory a little unnatural.

A popular proposal to get around this problem is supersymmetry, which extends the standard model via a symmetry between fermions and bosons. Supersymmetry would give no fewer than five different Higg-like particles. But so far there is no sign of supersymmetric particles in the LHC, causing some to doubt this solution (see Chapter 8). The light Higgs remains a puzzle. It could even spell disaster for our universe (see Chapter 7).

The dark Higgs

The Higgs boson, the particle credited with giving other particles mass, could also give rise to dark energy, the strange force that is causing the universe to fly apart faster and faster.

When the Higgs was discovered, researchers hoped to see anomalies in its behaviour that could lead them beyond the standard model of particle physics, which cannot account for phenomena such as dark matter. So far, it has been frustratingly normal, but perhaps its mere existence offers pathways to new physics.

The fields of the standard model create a certain density of energy that permeates the universe, one that would make it expand at an increasing rate. That energy density is much bigger than the value for dark energy we measure using observations of galaxies moving away from each other.

How could the Higgs solve this conundrum? Unlike other fields in the standard model, the Higgs field is scalar – it does not act in a specific direction. Even before the Higgs was discovered, physicist Lawrence Krauss of Arizona State University in Tempe was wondering whether other scalar fields existed that could interact with the Higgs field.

Krauss and James Dent, of the University of Louisiana at Lafayette, devised a new scalar field that does just that. The standard model says that the fields of all fundamental forces should merge at extremely high energies, meaning that there is also a unified, high-energy field out there. The new scalar field would have zero energy density, but it can use the Higgs to link up to this high-energy field, in the process acquiring energy of its own.

The precise amount is determined by a 'seesaw mechanism': if the energy of one field goes up, the energy of the other goes down. Since the unified field is so energetic, the new scalar field would be at very small energies. Krauss and Dent found that it would be of the right order of magnitude to account for dark energy.

FIGURE 3.5 Peter Higgs (1929–), scientist, in front of his portrait

Interview: The man behind the 'God particle'

Peter Higgs (see Figure 3.5) is Emeritus Professor of Physics at the University of Edinburgh, UK. In 1964, along with Robert Brout and François Englert at the Free University of Brussels, he proposed a new particle that would explain how other fundamental particles gain mass. In 2013 he co-won the Nobel Prize in Physics. New Scientist interviewed him a week after his eponymous particle was discovered in 2012.

Did the announcement of the discovery of the Higgs boson take you by surprise?

The week before all this started happening, I was at a physics summer school in Sicily. I didn't take any Swiss francs with me and my travel insurance policy expired the day I was supposed to fly back to Edinburgh. As the week went on, rumours began to fly, but it wasn't until the Saturday before the announcement that we knew for certain that something was up. We got a phone call from John Ellis, the former head of theoretical physics at CERN, saying, 'Tell Peter that if he doesn't come to CERN on Wednesday, he will very probably regret it.' I said, very well then, I'll go.

How were you feeling at that point?

I was getting excited. The final confirmation that the good news was coming came the evening before the CERN seminar. We had dinner at John Ellis's house and he cracked open a bottle of champagne.

It was obviously an emotional moment when the announcement was made at the seminar last Wednesday.

I was asked by a journalist at the seminar why I burst into tears after the presentation. During the talks I was still distancing myself from it all, but when the seminar ended, it was like being at a football match when the home team had won. There was a standing ovation for the people who gave the presentation, cheers and stamping. It was like being knocked over by a wave.

How did you celebrate?

With a can of London Pride ale on the flight back to London.

It was 48 years ago that you came up with a mechanism to account for the existence of mass, predicting the Higgs boson in the process. But one of your first papers on the subject was rejected, many of your peers thought your ideas were wrong at first, and Stephen Hawking bet against the discovery of the boson. Do you feel vindicated?

Yes, well, it's nice to be right sometimes. I didn't expect it to happen in my lifetime, at the beginning. Things began to change when the big colliders were built – LEP [the Large Electron–Positron Collider], the Tevatron and now the LHC. At the beginning no one knew what the mass of the Higgs would be, so it could have been too high to be discovered by these colliders.

Did you ever doubt that the particle existed?

No, I didn't really. It is so crucial for the consistency of the mechanism. You can remove the particle as a theoretical exercise, but then it becomes nonsense. I had faith in the theory behind the mechanism as other features of it were being verified in great detail at successive colliders.

It would have been very surprising if the remaining piece of the jigsaw wasn't there.

Several types of Higgs particle have been proposed, fitting various theories of particle physics. Which do you favour?

I'm a fan of supersymmetry, largely because it seems to be the only route by which gravity can be brought into the scheme. It's probably not even enough, but it's a way forward to get gravity involved. If you have supersymmetry, then there are more of these particles. That would be my favourite outcome.

You have always been a rather reluctant science celebrity. Is there a sense of relief and a hope that maybe now the attention will die down?

Relief is certainly part of it. It is still too recent for me to have come out of the upheavals of the last few days. The best I can hope for, I think, is some spells of quiet. At the moment, that's not looking likely. My inbox and doormat are full with emails and letters from people who want me to endorse their Higgs board game or to inaugurate the walkway of their new office atrium. There's even a microbrewery in Barcelona that wants to know what my favourite beer is so they can brew a similar one in my honour. It is quite mad.

What with your peers calling for you to get a knighthood and the Nobel Prize, there's no sign of a return to a quiet life just yet. Do you think much about what might happen next?

Well, come October when the prize is announced I shall probably suffer from what Nobel winner Sheldon Glashow called Nobelitis. You get jittery.

With all this attention, have you come up with a snappy one-liner to explain the Higgs mechanism yet?

No, I spend more time telling people that explanations by physicists who should know better are nonsense. The one that I object to is that the acquisition of mass by a particle is like dragging it through treacle. (That is a process where you are losing energy.) The trouble is that when I try to explain it in the way I would prefer, there are so many people that don't know the eighteenth-century physics that is needed. I explain it as being somewhat like the refraction of light through a medium.

The model I came up with in 1964 is just the invention of a rather strange sort of medium that looks the same in all directions and produces a kind of refraction that is a little bit more complicated than that of light in glass or water. This is a wave phenomenon but you can translate it into the language of particles by waving your hands and muttering the magical names of Einstein and de Broglie [who formulated the idea that waves could have particle properties, and vice versa].

Because several people came up with the mechanism at almost the same time, knowing what to call the particle has been a minefield. What are you calling it these days?

I can't see how it can continue to be called the Higgs boson. I reckon it will become just the H boson. Hopefully, in a particle physics context, it shouldn't get confused with hydrogen. I do sometimes still call it the Higgs boson so people know what I'm talking about. I don't call it 'the God particle'. I hope that phrase won't be used as much as it has been recently. I keep having to tell people that it's someone else's joke, not mine.

That label has made the particle sound very accessible, though.

That's true, but it has connotations that are simply misleading. It causes some people who don't know how the phrase arose to say rather foolish things. I've heard some people who have a background in theology try to make sense of it in terms of that. They don't understand it was just a joke; it was never meant to be taken seriously.

Why is the Higgs boson often called 'the God particle'?

To call the Higgs boson the 'God particle' is to invite the wrath of many a physicist. The term originates with Nobel laureate Leon Lederman, who led the charge to find the Higgs at Fermilab's particle accelerator and wrote a book about his search. According to Lederman, he wanted to call it 'that goddamn particle' but his publisher shortened it.

4
Quark tales

Most of you, by weight, is made up of quarks and gluons. These particles bind together in strange and complicated ways, but we can now calculate their behaviour with much greater precision than before — and so begin to answer some fundamental questions about our own existence.

Wrestling with quarks

Ask a science nut to name their heart's desire, and many might say: the answer to life, the universe and everything; failing that, a fully functioning lightsaber (see Figure 4.1). It is odd, then, that the scientific field that could conceivably provide both gets so little press. If we can really get to grips with the unglamorous proton and neutron, we can begin to explain how the material universe came to exist and persist, and explore mind-boggling technologies such as new types of laser and materials to store energy.

The main difference between protons and neutrons is that protons have a positive electrical charge, whereas neutrons are neutral. But they also differ slightly in mass. The neutron weighs in at 939.6 MeV and the proton at 938.3 MeV. That's a difference of just 0.14 per cent, but it matters. The extra mass

FIGURE 4.1 Lightsabers, the weapons of the Jedi, in action (from *Star Wars Episode V – The Empire Strikes Back*)

means that neutrons can decay into protons, but not vice versa. Stable protons can team up with negatively charged electrons to form robust, structured, electrically neutral atoms. If instead protons were heavier, they would decay into neutrons and the world would be a featureless neutron gloop.

The exact amount of the neutron's excess baggage matters, too. The simplest atom is hydrogen, which is a single proton plus an orbiting electron. Hydrogen was made in the Big Bang, before becoming fuel for nuclear fusion in the first stars, which forged most of the other chemical elements. All elements except hydrogen also contain neutrons. If the proton–neutron mass difference were just a little bigger, fusion would be more difficult, because there would be a bigger energy barrier to adding neutrons. The universe would be stuck at hydrogen.

However, if the mass difference were slightly smaller, hydrogen would have fused into helium in the early minutes of the cosmos, leaving little fuel to power long-lived stars like our Sun. Narrow the gap further, and hydrogen atoms would have collapsed: the proton would be able to eat its orbiting electron, turn into a neutron and spit out a neutrino. Then there would be no atoms whatsoever.

Where do they come from?

If proton and neutron didn't have these precise masses, we wouldn't exist. But the question of where the masses come from is fiendishly difficult to answer. We have known for half a century that protons and neutrons – collectively known as nucleons – are not fundamental particles, but made of smaller constituents called quarks. There are six types of quark: up, down, strange, charm, bottom and top. The proton has a composition of up-up-down, while the neutron is up-down-down.

Down quarks are slightly heavier than up quarks, but this does not explain the neutron's sliver of extra mass: both quark masses are tiny. It's hard to tell exactly how tiny, because quarks are never seen singly, but the up quark has a mass of something like 2 or 3 MeV and the down quark maybe twice that: just a tiny fraction of the total proton or neutron mass.

Like all fundamental particles, quarks acquire these masses through interactions with the all-pervasive Higgs field (the thing that the LHC shook up enough to create a detectable Higgs boson). But explaining the mass of matter made of multiple quarks clearly needs something else.

To reach the answer, we must scale the cliff face of quantum chromodynamics (QCD). Just as particles have an electrical charge that determines their response to the electromagnetic force, quarks carry one of three colour charges that explain their interactions via the strong nuclear force (see Chapter 2). QCD is the theory behind the strong force, and it is devilishly complex.

Quarks bind together to form matter such as protons and neutrons by exchanging gluons. Although gluons have no mass, they do have energy. That energy has a mass equivalent, according to $E = mc^2$. What's more, the exchange of gluons should also generate a sea of short-lived quark–antiquark pairs – virtual mesons – within each nucleon.

Quantum froth

To try to make sense of this quantum froth, over the past four decades particle theorists have invented and refined a technique known as lattice QCD. In much the same way that meteorologists and climate scientists attempt to simulate the swirling complexities of the Earth's atmosphere by reducing it to a three-dimensional grid of points spaced kilometres apart, lattice QCD reduces a nucleon's interior to a lattice of points in a simulated space-time tens of femtometres (10^{-15} m) across. Quarks sit at the vertices of this lattice, while gluons propagate along the edges. By summing up the interactions along all these edges, and seeing how they evolve step-wise in time, you begin to build up a picture of how the nucleon works as a whole.

Even with a modest number of lattice points (say 100 by 100 by 100, separated by one-tenth of a femtometre) that's an awful lot of interactions, so lattice QCD simulations require a huge amount of computing power. Complicating things further, quantum physics offers no certain outcomes, so these simulations must be run thousands of times to arrive at an average answer.

To work out where the proton and neutron masses come from, a group led by Zoltán Fodor of the University of Wuppertal, Germany, harnessed two IBM Blue Gene supercomputers and two suites of cluster-computing processors. In 2008 they finally arrived at a mass for both nucleons of 936 MeV, give or take 25 MeV. This confirmed that the interactions between quarks and gluons make up most of the mass of matter. You might feel solid, but in fact you're 99 per cent binding energy (see Figure 4.2).

These calculations were still not precise enough to pin down that all-important difference between the proton and neutron masses. What's more, the calculation omitted the effects of electrical charge. All the transient quarks and antiquarks inside the nucleon are electrically charged, giving them a self-energy that makes an additional contribution to their mass.

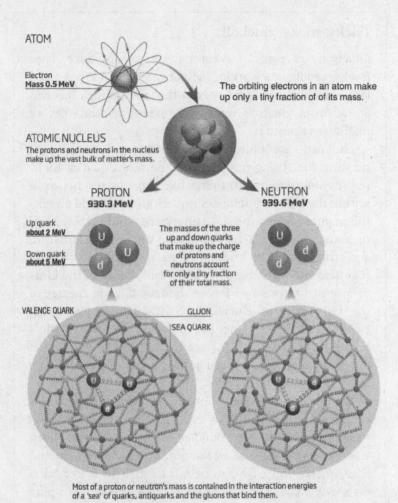

ATOM

Electron
Mass 0.5 MeV

The orbiting electrons in an atom make up only a tiny fraction of of its mass.

ATOMIC NUCLEUS
The protons and neutrons in the nucleus make up the vast bulk of matter's mass.

PROTON
938.3 MeV

NEUTRON
939.6 MeV

Up quark
about 2 MeV

Down quark
about 5 MeV

The masses of the three up and down quarks that make up the charge of protons and neutrons account for only a tiny fraction of their total mass.

VALENCE QUARK   GLUON

SEA QUARK

Most of a proton or neutron's mass is contained in the interaction energies of a 'sea' of quarks, antiquarks and the gluons that bind them.

FIGURE 4.2 Where matter gets its mass from

Tricky, sticky glueballs

Glueballs are particles consisting entirely of force. They might even make a working lightsaber. There's just one problem – they seem to exist only in theory. Although theorists are adamant glueballs must exist, experimentalists say we might never prove it.

Glueballs are bundles of gluons, the particles that transmit the strong nuclear force between quarks, sticking them together into things like protons and neutrons within the atomic nucleus. Simulations of a world chock-full of gluons suggest that an energy of around 1500 MeV, or about one-and-a-half times the energy contained in a proton, should be enough to stick a quantity of them together into a glueball. In 1995 Frank Close of the University of Oxford and fellow theorist Claude Amsler of the University of Zurich in Switzerland showed that two particles with energies of 1370 and 1500 MeV, which had just been discovered at CERN, might fit that bill. They have since been joined by a third candidate at 1710 MeV.

But the strong force is notoriously hard to do calculations with, and for simplicity's sake glueball simulations tend to assume a world with bundles of gluons and not much else. In the real universe, by the time you measure a glueball state, quarks will have also begun to stick to it like burrs on a sock, making it impossible to prove it ever was a pure glueball. The explanation for the three suggestive particles is most probably a substantial pot of glue contaminated by varying amounts of different quarks, says Close – and we'll probably have to content ourselves with that ambiguity.

A theorist's worst nightmare

The subtle roots of the proton–neutron mass difference lie in solving not just the equations of QCD but those of quantum electrodynamics (QED), which governs electromagnetic interactions. It is a theorist's worst nightmare to try to solve them both at the same time, but in 2014 Fodor and his colleagues developed a mathematical workaround to arrive at a value for the proton–neutron mass difference. The figure they came up with agrees with the measured value (although still with a fairly large uncertainty, about 20 per cent).

You might wonder what we gain by calculating from first principles numbers we already know. But quite apart from revealing why complex matter can exist, it means that we can now calculate many things that make the universe tick. Take the processes inside exploding stars. Supernova explosions were the first events to seed the universe with elements heavier than hydrogen and helium. With the new ability to marry QCD and QED, we might answer questions such as which heavy elements formed when.

The advance might also help clear up some of the problems surrounding fundamental physics. The Large Hadron Collider's discovery in 2012 of the Higgs boson, and nothing else so far, leaves many open questions. Why did matter win out over antimatter just after the Big Bang (see Chapter 5)? Why do the proton and electron charges mirror each other so perfectly when they are such different particles? In general, being able to make precise calculations with QCD could reveal more places where the standard model fails to explain experimental results, perhaps pointing us towards new fundamental physics.

It's also worth remembering that new technologies have often sprung from a deeper understanding of matter's workings. A century ago we were just getting to grips with the atom, an understanding on which innovations such as computers and lasers were built. Then came insights into the atomic nucleus, with all the technological positives and negatives those have brought, from power stations and bombs to cancer therapies.

Digging down into protons and neutrons means we might now find a deeper seam to mine. Gluons are far more excitable in their interactions with colour charge than are photons in electromagnetic interactions, so it could be that manipulating colour-charged particles yields vastly more energy than fiddling with things on the atomic scale. Frank Wilczek of the Massachusetts Institute of Technology – who won a share of the 2004 Nobel Prize in Physics for his part in developing QCD – has speculated that powerful new X-ray or gamma-ray sources might exploit sophisticated nuclear physics, for example.

Gluons, unlike photons, also interact with themselves, and so might possibly be able to confine each other into a writhing pillar of energy. This has led Wilczek to suggest, tongue in cheek, that gluons might make a *Star Wars*-style lightsaber. More likely is the prospect of better ways to harness and store energy. Nuclei can pack a lot of energy into a small space, says Wilczek, so accurate nuclear chemistry by calculation could lead to dense energy storage. Even that is probably still a long way off, but, with the new accuracy of QCD calculations, the road is at last open. Welcome to the quark age.

Pentaquarks: a long-sought new form of matter

Quarks are normally observed in sets of two or three, but groupings of four and five quarks were predicted in the 1960s. A tetraquark – made of four quarks – was sighted and confirmed in 2013, but pentaquarks remained concealed until 2015.

The pentaquark (see Figure 4.3) was observed at the LHCb experiment by studying the decay of particles called Lambda b baryons. This decay normally occurs in one smooth step, but the physics of the strong force means that there are sometimes intermediate states along the way. During one such not-so-smooth decay, a particle popped up with a signature that tallied with that of the theorized pentaquark. Since then, the discovery had been confirmed with more data.

Some physicists believe that the five quarks are tightly bound into a single well-defined particle, while others propose that it would be better described as a mini-molecule, consisting of a two-quark meson and a three-quark baryon held together by the strong force. Because the particle decays almost immediately, more experiments will be needed to determine its size and mass precisely.

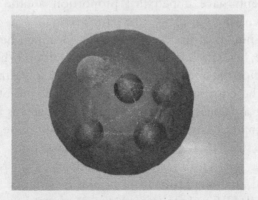

FIGURE 4.3 The possible layout of the quarks in a pentaquark. The five quarks might be tightly bonded. Or they might be assembled into a meson (one quark and one antiquark) and a baryon (three quarks), weakly bonded together.

The spin crisis

The proton's innards have been puzzling us since 1988, when researchers at CERN discovered that they could not account for the proton's spin. Spin is a quantum-mechanical property of a particle akin to a rotation about its own axis. Particles of different spins respond to magnetic fields in different ways, so it is a relatively easy thing to measure. The proton, for example, has a spin of ½. This spin must in some way come from the spin of the proton's components, just as the proton's one unit of positive charge comes from totting up the charge of the three valence quarks within it: two of charge +⅔ and one of charge −⅓.

By shooting protons apart with high-energy muons, CERN's European Muon Collaboration managed to measure the spin of the proton's interior quarks. They found it could account for only something like a quarter of the expected spin. Subsequent experiments have upped that proportion a little to around 30 per cent, but confirmed the basic result.

This 'spin crisis' has been the source of much head scratching since, because it means we don't understand the quantum structure of the proton. Ideally, we would solve the crisis by solving the equations of QCD, but these are monstrously difficult for a particle with as many moving parts as the proton. Such complications also overwhelm attempts to use supercomputers to simulate the origin of the proton's spin.

We are left with messy experiments to fill in the gaps. In 2014 and 2015 experiments using the Relativistic Heavy-Ion Collider (RHIC), at the Brookhaven National Laboratory, suggested that gluons might themselves carry a substantial proportion of the proton's spin, perhaps about 40 per cent.

Some of the proton's spin might have less to do with how quarks and gluons spin individually and more to do with how they orbit each other. As yet we have only the vaguest of ideas how we might go about measuring that. If there is such a thing as a theory of everything, we expect it will look a lot like QCD, only harder.

If we cannot understand what goes on within the humble proton, hopes will fade that we will ever be able to get to grips with that greater theory.

Why are heavy particles unstable?

A general rule of particle physics is that heavy things tend to decay, spitting out lighter particles, often with a lot of kinetic energy. The neutron decays into a proton (plus electron and neutrino). This is basic physics: things like to slip to states of lower potential energy, like a ball rolling down a hill. And as we know, mass is a form of energy. A lighter proton decaying into a heavier neutron would be like a ball spontaneously rolling uphill. For a given type of particle, more mass usually means a shorter lifetime; so the top quark, the W and Z particles and the Higgs are all very short-lived.

But why doesn't the proton decay into other light particles, say an electron and a positron? That's because of the underlying symmetries of particle physics. The symmetries tell us about physical quantities that are conserved, that cannot be changed (see 'Why symmetry rules the universe' in Chapter 2). The symmetries of the strong force say that baryon number must be conserved: you can't make a proton or a neutron out of nothing, or make one disappear. The proton is the lightest baryon, so there is no combination of particles that it can decay into, which makes it stable ... as far as we know (see Chapter 10).

5
Antimatter

The reality of antimatter is perhaps as extraordinary as anything in science fiction. It forms a whole shadow-world of particles, strangely scarce at least in our cosmic neighbourhood. Antimatter might never be a practical fuel for starships, but its explosive power could crack open the standard model and shine a light on why the universe is filled with matter.

Antimatter: a briefing

Every charged particle has an antiparticle with the same mass but the opposite electric charge. The proton has the negatively charged antiproton; the electron has the positively charged anti-electron, or positron. The possibility of antimatter first surfaced in equations formulated by British theoretical physicist Paul Dirac in 1928 – four years before American experimenter Carl Anderson found positrons in cosmic rays.

Many neutral particles have antiparticles, too. The neutron, for example, is made up of charged quarks. Turn these quarks into antiquarks by flipping their charges, and you have made an antineutron. Neutrinos don't have charged components, but in the conventional standard model they too have antiparticles – which spin in the opposite direction, and have the opposite lepton number. Matter and antimatter destroy each other, or annihilate, when they come into contact. An electron and a positron turn into two photons sent out in precisely opposite directions, each with an energy of 511 keV, corresponding to the mass of the electron (or positron).

The father of antimatter

Born in Bristol, UK, Paul Dirac (1902–84) made a crucial leap in our understanding of fundamental particles and forces. The equation he devised in 1928 predicted antimatter – an achievement for which he was awarded the 1933 Nobel Prize in Physics aged just 31. He was an eccentric and awkward character, with a particular love of Mickey Mouse; and to many he is considered to be the greatest British theorist since Isaac Newton. A stone inscribed with his famous equation describing the quantum behaviour of an electron can be found in the floor of London's Westminster Abbey.

Where is all the antimatter?

If you were to list the imperfections of the standard model, pretty high up would have to be its prediction that we don't exist. According to the theory, matter and antimatter should have been created in equal amounts at the Big Bang. They would have annihilated each other totally in the first second or so of the universe's existence, and the cosmos should be full of light and little else. And yet here we are. So, too, are planets, stars and galaxies. All, as far as we can see, are made exclusively out of matter.

There are two plausible solutions to this existential mystery. First, there might be some subtle difference in the physics of matter and antimatter that left the early universe with a surplus of matter. While the standard model predicts that the antimatter world is a perfect reflection of our own, experiments have already found suspicious scratches in the mirror. In 1998, CERN experiments showed that one particular exotic particle, the kaon, turned into its antiparticle slightly more often than the reverse happened, creating a tiny imbalance between the two.

That lead was followed up by experiments at accelerators in California and Japan, which in 2001 uncovered a similar, more pronounced asymmetry among heavier cousins of the kaons known as B mesons. The LHCb experiment at CERN is now using a 4,500-tonne detector to spy out billions of B mesons and pin down their secrets (see later in this chapter). But it won't necessarily provide the final word on where all that antimatter went. The effects seen so far seem too small to explain the large-scale asymmetry.

The second plausible answer to the matter mystery is that somehow, matter and antimatter managed to escape each other's fatal grasp. Somewhere out there, in some mirror region of the cosmos, antimatter is lurking and has coalesced into anti-stars, anti-galaxies and maybe anti-life.

When a hot magnet cools, individual atoms can force their neighbours to align with magnetic fields, creating domains of magnetism pointing in different directions. A similar thing could have happened as the universe cooled after the Big Bang, and those small differences could then have expanded into large separate regions over time.

These antimatter domains, if they exist, are certainly not nearby. Annihilation at the borders between areas of stars and anti-stars would produce an unmistakable signature of high-energy gamma rays. If a whole anti-galaxy were to collide with a regular galaxy, the resulting annihilation would be of unimaginably colossal proportions. We have not seen any such sign, but then again there is a lot of universe that we haven't looked at yet – and whole regions of it that are too far away ever to see.

Finding anti-helium or other anti-atoms heavier than hydrogen would be evidence for such an anti-cosmos. It would imply that anti-stars are cooking up anti-atoms through nuclear fusion, just as regular stars fuse normal atoms. On the International Space Station, an experiment called the Alpha Magnetic Spectrometer has been scouring cosmic rays for these anti-atoms since 2011. So far, we are still waiting for the first such emissary from the anti-cosmos.

Could we make an antimatter bomb?

The idea that humanity might one day harness the annihilative power of antimatter for destructive purposes has a ghastly fascination. It's the central idea in Dan Brown's novel *Angels and Demons* (2000), in which a bomb containing a quarter of a gram of antimatter threatens to obliterate the Vatican.

Relax, says Rolf Landua, a physicist at CERN. There's a very good reason why nothing like that is going to happen any time soon. 'If you add up all the antimatter we have made in more than 30 years of antimatter physics here at CERN, and if you were very generous, you might get 10 billionths of a gram,' he says. 'Even if that exploded on your fingertip it would be no more dangerous than lighting a match.' Patients undergoing PET scans have natural radioactive atoms in their blood-streams emitting tens of millions, if not more, positrons to no ill effect.

Even if physicists could make enough antimatter to build a viable bomb, the cost would be astronomical. A gram might cost a million billion dollars, Landua estimates. Frank Close, a particle physicist at the University of Oxford, points out that with current technology it would take us 10 billion years to assemble enough anti-stuff to make Dan Brown's bomb.

That is not to say we cannot harness antimatter in new ways. For decades, physicists have been able to make positronium – an 'atom' built from an electron and a positron. In 2007 physicists David Cassidy and Allen Mills of the University of California, Riverside, made the first molecules comprising more than one positronium atom. Positronium atoms quickly annihilate into high-energy gamma rays, so if you pack lots of them together, it should be possible to get them annihilating and emitting light in synchrony – creating a powerful gamma-ray annihilation laser, which might be used, for example, to set off nuclear fusion in reactors.

The Big Bang blip: solving the mystery of why matter exists

The banana on your kitchen counter is an embodiment of one of the universe's great mysteries, just waiting to be unpeeled. It is made of particles of matter, just like you, which is why you can see, feel and taste it. What you don't see is that 15 times a second or so it produces a particle of something else, something that vanishes almost instantaneously in a flash of light. That something else is antimatter.

Antimatter forms a whole mirror world of particles, identical in mass to those of normal matter but with opposite electrical charge. But it seems rather an afterthought. Around here, antimatter particles are produced only during interactions of high-energy cosmic rays in the atmosphere, or in radioactive decays such as those from the radioactive potassium-40 in bananas.

In one sense, that's unsurprising, given that antimatter and matter annihilate whenever they meet, giving out high-energy photons. But it leaves the mystery of how matter came to be so dominant.

So perhaps some blip at the beginning of the universe caused some matter to survive – and make everything from bananas to black holes, seahorses to stars. A difference of just one part in a billion between matter and antimatter would have been enough, because in the superhot soup of the Big Bang, particles and antiparticles would have repeatedly been created and annihilated, allowing the imbalance to build up.

Explaining matter's dominance

In the 1960s James Cronin and Val Fitch found that anti-kaons decay at different rates from their kaon counterparts. This phenomenon is known as charge parity (CP) violation. What this means is that, if you look at a particle reaction, and then the same reaction viewed in a mirror and with particles swapped for their antiparticles, you will see the two reactions proceeding at slightly different rates. It was just the sort of thing that might explain matter's dominance, but unfortunately the effects measured so far have been far too small to explain the dearth of antimatter in our universe.

Enter LHCb, a detector at CERN's Large Hadron Collider. The b stands for beauty, or bottom, depending on taste. Beauty quarks are much more massive than the up and down quarks, and very unstable, but LHC's proton collisions have enough energy to create them. They combine with other quarks to form B mesons, which could hold the key to matter's victory.

Experiments including BaBar at SLAC National Accelerator Laboratory in California and Belle at the KEK Laboratory in Japan have shown that CP violation occurs when particles containing b quarks decay, and the imbalance is greater than for kaons (see Figure 5.1).

Even so, decays of known particles so far account for only about one part in a trillion of the CP violation needed to explain the prevalence of matter around us. Meanwhile, the discovery of the Higgs boson has only deepened the mystery. Its mass of 125 gigaelectronvolts and observed decay rate fit predictions from the standard model of particle physics. If it had defied them, that might have pointed to the influence of particles as yet unknown, which might have provided another source of CP violation.

That throws the focus back on B mesons. The 'beauty factories' BaBar and Belle collided electrons and their antimatter equivalents, positrons, with energies in the range of billions of electronvolts – enough to produce only the lightest B mesons

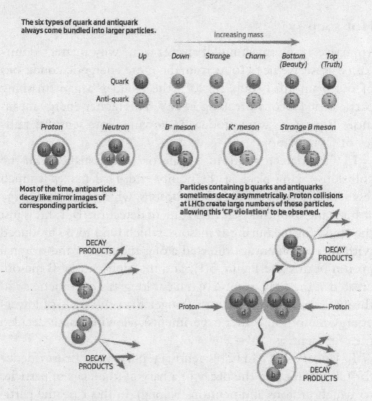

The six types of quark and antiquark always come bundled into larger particles.

Increasing mass

	Up	Down	Strange	Charm	Bottom (Beauty)	Top (Truth)
Quark	u	d	s	c	b	t
Anti-quark	ū	d̄	s̄	c̄	b̄	t̄

Proton   Neutron   B⁺ meson   K⁺ meson   Strange B meson

Most of the time, antiparticles decay like mirror images of corresponding particles.

Particles containing b quarks and antiquarks sometimes decay asymmetrically. Proton collisions at LHCb create large numbers of these particles, allowing this 'CP violation' to be observed.

DECAY PRODUCTS

Proton →   ← Proton

DECAY PRODUCTS

FIGURE 5.1 Subtle asymmetries in the way particles containing heavy quarks and antiquarks decay could hold clues as to why matter dominates antimatter in the universe.

in appreciable numbers. The LHC produces heavier B mesons than these experiments, in addition to light B mesons and B baryons. Looking at the broadest variety of beauty particle decay pathways, which should have different CP asymmetries, may lead to the best theory to describe CP violation in general.

Hot soup

To arrive at any meaningful answer as to why matter's dominance arose, we need to recreate the more energetic conditions of the primordial soup. In 2015 the collider began smashing particles at a record-breaking 14 TeV. This higher energy means more B particles were produced, allowing more sensitive studies of CP violation.

LHCb's detectors are in a cone or wedge shape, with the collisions taking place at the pointy end. That makes it much less sensitive to detecting Higgs bosons, which tend to emerge at high angles outside LHCb's zone of detection. But this is just the thing for measuring B mesons, which tend to be produced with their momentum directed along the line of the original proton beams (see Figure 5.1). And the high-energy B mesons created at the LHC can carve trails as long as a centimetre or so through the LHCb detectors (longer than the trails of lower-energy mesons in earlier experiments), allowing more detailed measurement.

In January 2017 LHCb scientists reported the first evidence for CP violation in the decay of a baryon (the class of particles to which protons and neutrons belong). In this case, the particle in question is the bottom Lambda baryon, which contains a bottom (beauty) quark. While the standard model predicts some degree of CP violation in the bottom Lambda baryon, analysis of more data from the LHC will reveal whether CP violation in this decay is more or less than expected. If it disagrees with the standard model, that would suggest new physics relevant to the fundamental question of why there is more matter than antimatter.

Data taken before the LHC's upgrade showed hints of deviations from standard-model predictions in the rare decay of a

B meson to a kaon and two muons. Since then, with more precise LHCb results, the differences with respect to the theoretical model have got even larger. Still, the theoretical prediction itself has been challenged, and the particle physics community does not agree that it's totally solid.

Additionally, there are hints of an imbalance between two decays – that of a B meson to a kaon, an electron and a positron; and that of a B meson to a kaon and two muons. But, again, the evidence isn't yet strong enough to tell whether some unknown physics is responsible for the anomaly or whether it can help to explain the matter mystery.

Another focus is the decay of the strange B meson, made from a b-antiquark and a strange quark. This can spontaneously transform into its antiparticle (b-quark plus antistrange quark), and looking at whether the reverse process happens in the same way provides another avenue to study CP violation. Also, three times in every billion, it decays into a muon and an antimuon. This decay may be rare, but its end state is very easy to see because muons leave a trail right through the detector all the way to the outermost layer (see Figure 5.2). That makes it a 'golden channel' to search for new physics, says Tara Shears of the University of Liverpool, UK, who works on the LHCb experiment.

Here, the interest may not just be in CP violation but also in hints of phenomena predicted by theories such as supersymmetry that posit the existence of new massive particles. Rare B meson decays should be especially susceptible to the influence of unseen massive particles, perhaps giving the LHCb experiment a sneaky way to prove their existence without detecting them directly. Any deviation from B mesons' expected rates of decay could mean that hidden particles are participating.

LHCb Event Display

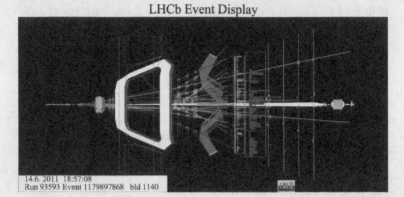

14.6. 2011 18:57:08
Run 93593 Event 1179897868 bld 1140

FIGURE 5.2 LHCb collisions create hordes of particles and antiparticles.

The strange B meson decay is so rare that experiments to date have seen very few examples, and so far its behaviour has not revolutionized the standard model, nor have other B mesons at the LHC. But more observations of beauty particles with the revamped particle accelerator could make all the difference.

Penguin clues

In the search for differences between matter and antimatter, one type of anomaly could finally be telling us something. The heavier types of B meson decay into tau particles more often than the standard model predicts. Such decays are known as penguin processes, because when physicists sketch one out as a Feynman diagram, they end up with something that looks like a penguin.

Unbalanced penguins have now popped up in three different experiments, including LHCb. The energies and trajectories of muons produced in B meson decays are also persistently aberrant in LHCb data. These anomalies may be an indication of new physics. Perhaps B mesons decay into as yet undetected particles before decaying further into muons, skewing their final distributions? Theorists speculate that the culprit could be a leptoquark – a hybrid of lepton and quark-type particles – or a new kind of Higgs boson.

Although these observed imbalances are small, they might be an indication of a path to follow. At higher energies, similar penguin processes might lead to greater imbalances.

Antigravity: does antimatter fall upwards?

If you took a piece of antimatter and released it, with free passage through the hostile world of matter, is there a chance that it would magically float up (see Figure 5.3)? Few physicists would bet on it but, until we do the experiments, we simply don't know.

Scepticism about antigravity dates back to the 1950s, when the physicist Hermann Bondi was pondering the implications of general relativity, Einstein's theory of how gravity arises from warping the fabric of the universe. Gravity is an odd sort of force, not least because it only ever works one way. With electromagnetism, say, there are positive and negative charges that attract and repel. With gravity, however, there are only positive masses that always attract.

FIGURE 5.3 What goes up … The gravity-defying artwork of Li Wei
(Beijing, China, 2010)

Bondi showed what a bizarre world it would be if that were not the case, demonstrating that negative mass would end up pursuing positive mass across the universe ever faster. This sort of runaway motion does not appear to exist – but we should be careful about what conclusions we draw, says Sabine Hossenfelder of the Frankfurt Institute for Advanced Studies in Germany. It may be that, rather than discarding negative mass, we need to modify general relativity.

There are two types of mass: gravitational mass quantifies how strongly an object feels the force of gravity, while inertial mass quantifies an object's resistance to acceleration. Many experiments have shown that these two quantities always have the same value – a mysterious equivalence that lies at the heart of Einstein's description of gravity, general relativity. But all these experiments have involved normal matter. Could antimatter break the equivalence principle by forming an object with normal inertial mass, but a negative gravitational mass?

Another corner of the universe

In a few labs around the world, the search for negative mass goes on. Antimatter is a promising place to look. It is just like normal matter but with the opposite electric charge and a few other mirrored quantum properties. That doesn't mean it should have opposite mass too, but if it did, that might help with another mystery: where all the antimatter went. A repulsive gravitational interaction could have driven matter and antimatter away from each other so they never had the chance to annihilate in the early universe. Since then, the ongoing expansion of the universe would have driven the two farther apart – and the antimatter might eventually have created its own galaxies in other corners of the universe.

With the technological possibilities that levitating matter away from Earth's surface might bring, even the US air force wants in. It has given millions of dollars to antimatter researchers over the years. Unfortunately, doing the experiments turns out to be quite a task. To start, you need a home for antimatter that is almost entirely free of normal matter. That requires some of the emptiest boxes on Earth, containing just a few hundred gas molecules per litre. Then, to stop the antimatter banging into the sides of the box and instantly annihilating, you must slow it down by cooling it to within a few degrees of absolute zero and then catch it in a trap of electromagnetic fields.

Six experiments competing to measure antimatter's fundamental properties are housed in CERN's vast Antimatter Deceleration Hall. Below, a beam of particles from CERN's Proton Synchrotron accelerator smashes into a block of metal, creating a plethora of particles. A system of magnets selects the antiprotons and funnels them into a ring of more magnets that keep them on course as they are decelerated for trapping.

Close to identical

Experiments have been running here since the 1990s, studying whether antimatter and matter particles truly are as close to identical as we think. In 2015, by measuring how antiprotons danced around in a magnetic enclosure known as a Penning trap, the Baryon Antibaryon Symmetry Experiment (BASE) produced the most precise measurement yet of the ratio between their inertial mass and their charge. The result showed that it is the same as a proton's, to about 69 parts per trillion, four times more precise than the previous best value.

In November 2016 the neighbouring ASACUSA experiment produced the most accurate measurement yet of the antiproton's inertial mass, finding no evidence of a different value from the proton's.

But is the mass positive or negative in its gravitational effects? That question takes the experiments to a new level of detail. Gravity is weak and easily overwhelmed by the electromagnetic force, so using charged particles such as antiprotons and controlling them with magnetic fields won't do. You could try getting an antiproton in position and shutting off the magnets to see which way it falls, but the antimatter's electrostatic interactions with its surroundings would overwhelm any gravitational push or pull it might feel.

A better bet is neutral atoms of antimatter, such as antihydrogen (see 'Anti-elements', below). Their electrostatic interaction is not strong enough to swamp gravity, but very strong magnetic fields will still hold them in place. CERN's Antihydrogen Laser Physics Apparatus (ALPHA) experiment routinely traps and holds bunches of antihydrogen atoms for about 15 minutes.

In 2013 ALPHA published a proof of principle measurement, briefly collecting a cloud of 434 anti-atoms, turning off the magnets and tracking their subsequent motion by where they annihilated. It was a crude test, and inconclusive – the final answer was compatible with the antiparticles having either negative or positive gravitational mass. Work on a more complex version that gives the particles more space to fall should start in 2017. Getting the necessary accuracy will not be easy, because the anti-atoms ALPHA uses are relatively hot and so vibrate, which clouds the issue. But large enough numbers of anti-atoms should help us answer the central question.

A further CERN experiment, AEGIS, also aims to perform tests within a few years. Daniel Kaplan of the Illinois Institute of Technology in Chicago is planning experiments with muons, heavier cousins of the electron, while a team led by David Cassidy of University College London is planning to use positronium.

Back at CERN, the Gravitational Behaviour of Antimatter at Rest, or GBAR, experiment aims to tackle the question using a single antihydrogen ion, a combination of one antiproton and two positrons. In theory, it should be easy to hold this charged speck in place with magnetic fields and cool it with lasers. The idea is then to knock off a positron using another laser, making the anti-atom neutral. At this point it would cease to feel the effect of the trapping field and fall – up or down. GBAR's head, Patrice Perez, says they expect to make measurements sensitive to detect even a 1 per cent deviation from the gravity felt by normal matter.

Experimenters do not expect antimatter to fall up, but if it falls at all differently, that would still be hugely interesting. It could imply the existence of forces that modify gravity, whose effects cancel out on normal matter but not on antimatter. Similar gravity-modifying effects might be produced if the graviton, a quantum particle proposed to carry the force of gravity, has a small mass, rather than being massless as is usually assumed.

Anti-elements

Two CERN experiments, ATRAP and ALPHA, are already making antihydrogen – the simplest anti-atom possible, just an antiproton and a positron bound together. Their aim is to produce enough of it and hold it for long enough to compare the spectrum of light it emits with that of regular hydrogen. Even the slightest difference between the two would shake up the standard model, which predicts they should be exactly the same.

In 2016, by creating and holding just 14 atoms of antihydrogen, Jeffrey Hangst and his colleagues in the ALPHA collaboration managed to see an antimatter spectral line for the first time. They found that the lowest-energy photons in the antihydrogen spectrum do have the same wavelength as those of hydrogen.

Can we ever expect physicists to make more complex antimatter – antihelium, organic antimolecules made from anticarbon and a whole anti-periodic table? The problem is that every anti-atom has to be built one

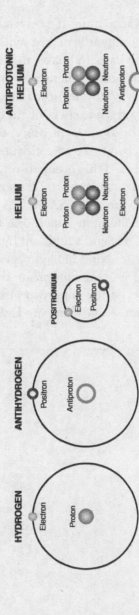

FIGURE 5.4 Anti-engineering: physicists have managed to make mirror-image hydrogen, an electron–positron pairing called positronium, and helium with an antimatter twist.

subatomic antiparticle at a time. If you want to make antideuterium – like antihydrogen but with an added antineutron – you first have to make the antineutron. Antineutrons are neutral, making them impossible to steer in the conventional way with electromagnetic fields, so you just have to make great numbers of them and hope that, for every million or so antineutrons you make, one ends up in the right place to make an antideuterium atom.

While no one has cracked that problem yet, one experiment at CERN is using a neat short cut to make something other than antihydrogen. ASACUSA has created atoms of antiprotonic helium, in which one of the electrons orbiting a helium nucleus is replaced by an antiproton (see Figure 5.4). By studying the light spectra emitted by this composite matter–antimatter atom, the electrical and magnetic properties of the antiproton can be measured with great precision – and compared with those of an ordinary proton.

6

The little neutral ones

Neutrinos are the most elusive pieces of matter. These ghostly particles pass straight through the Earth virtually unhindered. Although neutrinos are part of the standard model, their shape-shifting powers hint at something beyond it.

What exactly are neutrinos?

With a neutral charge and nearly zero mass, neutrinos are the shadiest of particles, rarely interacting with ordinary matter and slipping through our bodies, buildings and the Earth at a rate of trillions per second. First predicted in 1930 by the Austrian-born physicist Wolfgang Pauli, who won a Nobel Prize for this work in 1945, they are produced in various nuclear reactions: fusion, which powers the Sun; fission, harnessed by humans to make weapons and energy; and natural radioactive decay inside the Earth.

How do we know they are there at all? Wily neutrinos usually avoid contact with matter, but every so often they crash into an atom and produce a signal that allows us to observe them. Fredrick Reines first detected them in 1956, garnering himself a Nobel Prize in 1995. Most commonly, experiments use large pools of water or oil. When neutrinos interact with electrons or nuclei of those water or oil molecules, they give off a flash of light that sensors can detect.

A lot of expense and extreme engineering go into detectors that are sunk into the ground to shield them from extraneous particles that might interfere with them. For instance, OPERA lies inside the Gran Sasso Mountain in Italy. This works because neutrinos shoot straight through rock. Other detectors use the whole planet as a shield, pointing their detectors downwards. One such, ANTARES, is under the Mediterranean Sea, while another, IceCube, is buried in Antarctic ice.

The stealth of neutrinos belies their potential importance. Take extra dimensions. Most particles come in two varieties: ones that spin clockwise and ones that spin anticlockwise. Neutrinos are the only particles that seem just to spin anticlockwise.

Some theorists say that this is evidence for extra dimensions, which could host the 'missing' right-handed neutrinos.

Just like the charged leptons, neutrinos come in three flavours: electron, muon and tau. But, unlike charged leptons and any other particles we know of, they can morph from one flavour to another.

Neutrinos: the next big small thing?

'Uneasy lies the head that wears a crown,' wrote Shakespeare. The same could be said today of the standard model of particle physics, our most successful description of the building blocks of matter and their interactions. The discovery of the Higgs boson stands as the theory's crowning achievement, validating a prediction made nearly four decades ago and filling the model's last major gap. Yet we are as eager as ever to knock it from its throne, to discover the new physics that must surely supersede it.

Could neutrinos now guide us to this goal? Despite its power, the standard model leaves many questions unanswered – questions such as the nature of dark matter – the mysterious, invisible material thought to make up more than 80 per cent of the mass of the universe. Then there is dark energy, the stuff reckoned to be causing the universe's expansion to accelerate. In what must rank as our worst prediction ever, quantum physics overestimates dark energy's magnitude by a factor of 10^{120}. The standard model cannot explain how matter survived annihilation with antimatter in the Big Bang, or how gravity fits into the picture. It is riddled with free parameters, troublingly arbitrary numbers that have to be fed into the theory by hand, for example to set the strength of each force.

Researchers had hoped that the Higgs would lead to new physics that could answer these questions. But with the Higgs behaving largely as expected so far, the real key to the kingdom beyond the standard model may lie with the neutrino – a different sort of particle. Ghostly, mysterious and antisocial, neutrinos rarely deign to interact with the world of common matter around them, since much of what is known about them lies outside the standard model.

Minuscule mass

The three neutrinos we know about are paired with the electron and its two heavier cousins, the muon and the tau. A trio of antineutrinos also exists, paired with the positively charged antiparticles of the electron, muon and tau to complete the extended lepton family (see Figure 2.2). From the outset, the standard model wrongly assumed that neutrinos have no mass. But we now know that they do have mass – although minuscule – which is behind their uncanny ability to shape-shift from one type into another. Many new theories hope to fill this gap in understanding, including grand unified theories, which aim to unite all the forces of nature (except gravity) as well as supersymmetry and string theory.

Despite their aloof nature, neutrinos have a long history as problem-solving particles. Physicist Wolfgang Pauli conceived of them in 1930 in order to conserve energy and momentum in radioactive beta decays, and since then they have nestled neatly into the standard model's orderly picture of particles.

But a crack in the standard model description of neutrinos came in 1998, from results at the Super-Kamiokande experiment in Japan. Neutrinos are emitted or absorbed with electron, muon or tau flavour, like single scoops of ice cream.

Super-Kamiokande studied muon neutrinos from cosmic rays striking the atmosphere and found that some of them morphed into electron neutrinos on their way through the Earth.

There had been earlier hints of this oscillation between flavours, but this was clear evidence – and since then, experiments investigating neutrinos created in nuclear reactors, particle accelerators and nuclear decay processes in the Sun have confirmed that, however they start out, neutrinos shape-shift into a mixture of all three flavours on their journey. And the only way such morphing can happen is if neutrinos have some mass.

The propagation of flavours

This is a very quantum phenomenon. In quantum mechanics, one particle can be in a superposition of more than one state – and for neutrinos, it turns out, a state with a well-defined flavour is a blend of three different states with a well-defined mass. The three mass states move at slightly different speeds just below the speed of light, and it's their precise line-up that determines what flavour is detected when a neutrino is observed. So the three neutrino flavours propagate through space as a constantly changing mixture.

Some grand unified theories, predict neutrinos with mass. Pinning down the precise masses could tell theorists which theory to pursue.

Measuring the mass of a particle that can sail unhindered through a light year of lead is easier said than done, but one way could be measuring radioactive beta decays. In a typical beta decay, a neutron inside an atomic nucleus turns into a proton while spitting out an electron and an electron antineutrino. In theory, the antineutrino's mass could be inferred from the energy and momentum of the accompanying electron. Neutrinos are so light that it has been impossible to achieve the sensitivity needed, but a sensitive experiment being built at the Karlsruhe Institute of Technology in Germany, called KATRIN, may yet manage it.

Meanwhile, the tightest limits on neutrino mass come from the cosmos, because these particles affect the mix of elements created in the Big Bang and supernovae, the expansion rate of the universe, the cosmic microwave background (CMB), and how matter coalesced into galaxies and galaxy clusters. A combination of the most recent cosmological measurements, including observations of the CMB from the Planck space observatory, reveal that the sum of the three neutrino masses cannot be more than about 0.13 electronvolts (eV) – less than a millionth that of the electron.

Because of their constant shape-shifting, it is difficult to break that sum into the masses of the three kinds of neutrino. Researchers are gradually refining their measurements of the masses and the mixtures of mass states that define the neutrino flavours.

Any theory that explains neutrino mass has to explain why it is so vanishingly small compared with any other particle. One theory suggests that the three known neutrinos may be shadowed by one or more 'sterile' neutrinos that feel only the force of gravity. Through a process called the seesaw mechanism, these heavy, invisible neutrinos suppress the masses of the detectable ones.

All of this makes these feather-light, antisocial shape-shifting particles even more enigmatic. They are hiding something, but what it is and what it will tell us about physics beyond the standard model remain to be seen.

The amazing disappearing faster-than-light neutrino

In September 2011 the OPERA collaboration for detecting tau neutrinos shocked the world. It made an announcement that neutrinos zipping from CERN in Switzerland to detectors beneath the Gran Sasso Mountains in Italy were outpacing the speed of light, a feat that violated Einstein's rules of relativity and opened the door to exotic physics.

But over the next few months, two errors – a leaky fibre-optic cable and a malfunctioning clock – emerged, which slowed the neutrinos back down. We now know that neutrinos travel at almost exactly the speed of light. 'Although this result isn't as exciting as some would have liked, it is what we all expected deep down,' said then CERN research director Sergio Bertolucci.

Making antimatter matter

A hunt for the rarest of events could reveal why matter dominates the universe – if we can spot something that happens once every 100 trillion trillion years. You could spend an eternity waiting and watching, and still never see it happen. And yet thousands of physicists are mounting a search for this fabulously rare process, a form of radioactive decay that, if sighted, could reveal why the universe contains anything at all.

Radioactive decay is nature's alchemy. It is capable of transforming certain heavier elements into lighter ones, but it runs on its own schedule – some elements have lifetimes of minutes, others millennia. These radioactive processes are vital to our existence, with beta decays, for example, helping to power the Sun. The most familiar type of beta decay causes a neutron in an atomic nucleus to transform into a proton, ejecting an electron and an antineutrino, the neutrino's antimatter partner, in the process.

In 1935 physicists predicted that certain nuclei might undergo two such beta decays at once. This rarest known form of nuclear decay is only likely to occur in a given nucleus once every 10^{19} to 10^{24} years. Keep watching a large enough collection of atoms, however, and you improve your chances of seeing one happen, with the result that we have now spotted it in 11 different heavy nuclei.

Even the rarity of these decays has nothing on what researchers are looking for now. Neutrinoless double beta decay involves two neutrons transforming into two protons and two electrons – without producing any antineutrinos at all (see Figure 6.1).

For this vanishing trick to occur, something remarkable needs to happen. The two antineutrinos need to effectively cancel each other out, much as particles and their antiparticles mutually annihilate when they come into contact. If these two identical particles are to neutralize each other, however, neutrinos

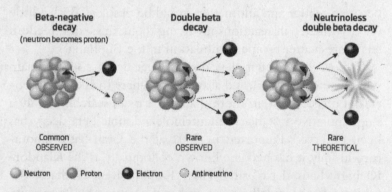

FIGURE 6.1 All radioactive beta decays are expected to produce some kind of neutrino. If the neutrino is its own antiparticle, however, one in a hundred trillion trillion decays will result in no neutrinos at all.

and their antiparticles must be one and the same – they must be both matter and antimatter at the same time.

Spotted in the wild

Even though neutrinos were spotted in the wild for the first time in 1956, there is still much we don't know about them. Part of the problem is how little notice they take of everything else in the universe – billions pass through you every second from the Sun, and they would emerge unscathed on the other side of a light year of lead. Even the seemingly simple question of whether or not neutrinos have mass was only resolved in the early 2000s, a result deemed so significant that it won 2015's physics Nobel Prize. But we still don't know why their masses are so small.

In 1937 the Italian physicist Ettore Majorana predicted that mass-bearing neutrinos would possess an intriguing property. As the only fundamental particle of matter with no electric charge, it would theoretically be possible for them to be their own antiparticles. For such 'Majorana neutrinos' the distinction between matter and antimatter would be obsolete. With a little CP asymmetry in reactions involving them, this extra freedom lets more matter escape annihilation in the Big Bang.

Discovering neutrinoless double beta decay could explain where we come from, and also hint at where to take our theories of physics. But, in more than 50 years of searching, only a single putative instance of neutrinoless double beta decay has been observed. Reported in 2001 at the Gran Sasso laboratory in Italy, it has become known colloquially as the Klapdor-Kleingrothaus decay after Hans Klapdor-Kleingrothaus, the University of Heidelberg researcher who stood by its veracity for over a decade. Most physicists remained unconvinced, but the result tantalized anyone with an interest in the decay.

The drive to confirm or rule out the Klapdor-Kleingrothaus event led to a burst of activity, inspiring the construction of experiments at Gran Sasso, in New Mexico and in Japan. Since 2015 they seem to have conclusively ruled out that event, since if it had been real they should have seen many more.

Could we communicate using neutrino beams?

We could and we have. The first neutrino-borne message was transmitted in 2012, when a team at Fermilab in Batavia, Illinois, used the Neutrinos at the Main Injector (NuMI) beam to fire pulses containing trillions of neutrinos towards the MINERvA detector, 1 kilometre away. The team encoded the word 'neutrino' using a standard binary communications code. Because neutrinos are so hard to detect, it took 142 minutes of repeated transmissions to make the message clear at the other end.

Neutrinos rarely interact with other forms of matter, so they pass through most objects unimpeded – even the Earth's core. That makes them potentially useful as messengers, perhaps to communicate with hidden submarines. Even a very low-bandwidth system might be useful for exchanging encryption keys.

A very long wait

These larger experiments have also refined our measurement of the decay lifetime, extending the minimum time you would expect to wait to see an individual nucleus disintegrate via neutrinoless double beta decay to 10^{25} years.

Now, at least eight new or revamped experiments are looking to spot the process, and these detectors just might be big enough and sensitive enough to achieve this (see Figure 6.2). They all operate along roughly the same lines: amass quantities of an extremely pure isotope deep underground, where it is shielded from the bombardment of cosmic particles that could swamp a detection. Then, be patient. Rely on the law of large numbers for a decay eventually to occur.

At present, only a handful of isotopes are both sensitive and abundant enough to be useful for experiments of this scale. Each has its merits, but all eyes are currently on germanium. Its concentrated crystalline form results in a more compact apparatus that can measure the energy of the two emitted electrons very precisely, making it easier for researchers to distinguish true decays from background events. GERDA, the Gran Sasso experiment that helped rule out the Klapdor-Kleingrothaus decay, is using 40 kilograms of germanium. Its main rival, the Majorana Demonstrator, located in a former gold mine deep beneath the town of Lead in South Dakota, has been taking data with a 44-kilogram device.

However, based on our most optimistic estimates for the decay lifetime, we need at least a tonne of detection material to be sure of getting a signal. This is approaching the limit of what a single experiment can supply, meaning that researchers will need to pool resources. Those working on Majorana and GERDA have discussed joining forces to build the first experiment using over a tonne of optimized isotope. That requires at least 10 tonnes of raw germanium that can be purified – almost 10 per cent of the annual global supply, which would otherwise go to technology companies to use for building faster computer chips.

An undiscovered particle?

Even if physicists see such decay, it won't be enough to prove that Majorana neutrinos are responsible. More complex mechanisms could also be causing this rare decay, with some undiscovered

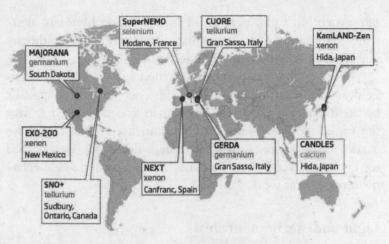

FIGURE 6.2 Detecting neutrinoless double beta decay requires isolating large quantities of the right radioactive substance deep underground and waiting for one to happen. Many experiments are under way across the world, using a variety of isotopes.

particle taking the role of the neutrinos. Distinguishing between such mechanisms is not easy, but new experiments are soon to add to physicists' arsenal – including KATRIN, under construction in Germany, which aims to establish the masses of the three fundamental neutrinos once and for all, and DUNE in the US, which will determine which mass goes with which flavour. If neutrinoless double beta decay is indeed caused by Majorana neutrinos, physicists can use the data from these experiments to calculate a lifetime for the process. They can then see whether these results match those from direct searches.

If they don't match, that would mean something else causes neutrinoless double beta decay. Such mechanisms are predicted by a host of theories now being searched for directly at CERN's Large Hadron Collider, including some variants of supersymmetry. Yet, especially in physics, the simplest solutions often turn out to be best – and another thing Majorana neutrinos have going for them is that they could settle the thorny questions of why neutrinos are so light.

Most fundamental particles get their masses via the Higgs field, an all-pervading field whose existence was proved in 2012 by the discovery of the Higgs boson, its associated particle. But the vanishingly low masses of the neutrinos suggest that these particles barely interact with the Higgs field at all. That has led some physicists to believe that another mass-generating mechanism must be at work.

Light and sterile neutrinos

The most popular of these is the seesaw mechanism in which heavier sterile neutrinos that out far outweigh any of the other fundamental particles bring down the masses of their lighter cousins.

The sterile neutrinos would have decayed in the first instants of the universe, leaving only the lighter variety behind. It's a mathematically elegant solution that is predicted by most grand unified theories – but it only works if the neutrino is a Majorana particle. In other words, there cannot be a separate and distinct antineutrino.

Sterile neutrinos, too, are under siege from experiment. Recent results from the IceCube neutrino detector at the South Pole (see Figure 6.3) have ruled out the existence of sterile neutrinos within a certain mass range, although heavier ones may still exist.

Whether the next generation of detectors strengthens the case for a Majorana neutrino or not, they will bring us a step closer to understanding the weird and wonderful processes neutrinos are involved in. Let's just hope we don't have to wait until the end of the universe to find it.

FIGURE 6.3 The IceCube neutrino observatory at the South Pole. Antarctica is a fertile hunting ground for neutrino oscillations.

Neutrinos hint at why antimatter didn't blow up the universe

A pair of experiments designed to study the behaviour of neutrinos could be giving us a sign of why matter beats antimatter in the early universe.

The T2K experiment in Japan watches for oscillations between electron, muon and tau neutrinos as the particles travel between the J-PARC accelerator in Tokai and the Super-Kamiokande neutrino detector in Kamioka, 295 kilometres away. In 2013 the team announced they had seen muon neutrinos becoming electron neutrinos by the time they reached Super-Kamiokande. They then ran the experiment with antineutrinos, to see whether there was a difference between how the ordinary particles and their antimatter counterparts oscillate. In July 2016 Hirohisa Tanaka of the University of Toronto, Canada, reported that the antimatter versions seemed less apt to alter: fewer muon antineutrinos become anti-electron flavour.

On its own, the T2K result does not have strong statistical significance, but it is backed up by similar results from NoVA, an experiment that sends neutrinos between Illinois and Minnesota. If the results are confirmed, this lopsidedness could be a sign of why matter prevailed 13.7 billion years ago.

The world's mightiest neutrino detectors

Some heavyweight technology has been used to study the elusive neutrino – in Japan, Canada, the USA, Germany and China:

- **Super-Kamiokande, Japan**
  Neutrinos interact so rarely with matter that vast experiments are needed to spot them. Central to the Super-Kamiokande experiment in Japan is a huge stainless-steel tank, 39 metres in diameter. It is filled with 50,000 tonnes of purified water. On the rare occasion that a neutrino interacts with the water, it creates a charged particle that then produces a flash of light. More than 13,000 sensitive light detectors surrounding the tank watch for these flashes. Super-Kamiokande showed that neutrinos change from one type to others as they travel, like strawberry milkshake turning into chocolate or vanilla.

- **Sudbury Neutrino Observatory (SNO), Canada**
  SNO is located more than 2 kilometres underground in a nickel mine in Ontario, Canada. Like Super-Kamiokande, its deep underground location shields it from cosmic rays.

Before the experiment was turned off in 2006, the SNO vessel was filled with 1,000 tonnes of heavy water, and spotted neutrinos in a similar way to Super-Kamiokande. SNO specialized in spotting neutrinos coming from the Sun. The experiment's director, Arthur McDonald, co-won the 2015 Nobel Prize in Physics for discoveries made at SNO showing that neutrinos have mass. Plans are under way to upgrade the experiment.

- **The Liquid Scintillator Neutrino Detector (LSND), USA**
  The LSND has provided neutrino physics with its most puzzling results. The experiment ran at the Los Alamos National Laboratory in New Mexico until 1998. It studied neutrinos created as a by-product of particle collisions and looked for them switching back and forth between three types. Yet the results didn't match expectations: it is as if the neutrinos vanish. The results can be explained if a fourth type of non-interacting, sterile neutrino exists. Later experiments appeared to confirm LSND's findings. However, an analysis of almost 100,000 neutrino events at the IceCube Neutrino Observatory at the South Pole casts some doubt on the existence of sterile neutrinos.

- **KATRIN, Germany**
  Neutrinos were thought to be massless, but their ability to change from one type to another shows that this cannot be true: neutrinos do have mass, albeit puzzlingly low. For many years, researchers have tried to measure this mass but their equipment has not been sensitive enough. Perhaps the KATRIN experiment will be. Based in Karlsruhe, Germany, it will look at the beta decay of tritium, which produces a neutrino, an electron and a helium-3 nucleus. Although the neutrino escapes detection, its mass can be

calculated from the energy and momentum of the electron. Tests at the experiment began in 2016, and measurements are due in 2017.

- **Daya Bay Reactor Neutrino Experiment, China**
  Located 50 kilometres north of Hong Kong, the Daya Bay experiment studies the antineutrinos flooding out from two nearby nuclear reactors. The team published their first results in 2012, completing the picture of how neutrinos and antineutrinos change flavour. Its unique design has paved the way for future experiments such as the much larger Jiangmen Underground Neutrino Observatory (JUNO) in Kaiping, China. Work began on the construction of JUNO in 2015, with the first results expected in 2020.

Interview: Searching for neutrinos from deep space

In 2014 Ray Jayawardhana, Professor of Astrophysics at the University of Toronto, Canada, told New Scientist *how the search for neutrinos from deep space gives us a new vista on some of the most violent processes in the universe.*

What's so interesting about neutrinos?

They are elementary particles with rather quirky properties. They hardly ever interact with matter, and that makes them really difficult to pin down. Trillions pass through your body every second but there's only maybe a 25 per cent chance that one will interact with an atom in your body in your whole lifetime.

Where do they come from?

Some come from the heart of the Sun; others are produced in the upper atmosphere when cosmic rays hit atoms. Then there are geoneutrinos that are produced in the Earth's interior as radioactive elements decay. The vast majority of neutrinos that pass through Earth are from those three sources. But there's a great deal of interest in detecting neutrinos that come from much farther away – cosmic neutrinos.

Why are cosmic neutrinos such a big deal?

Some of the more violent phenomena in the universe produce neutrinos. So there are some really fundamental questions that cosmic neutrinos allow us to probe. So far, though, only two batches have been detected. The first were from the supernova 1987A, a star that exploded in a satellite galaxy of the Milky Way. More recently, the Ice-Cube Neutrino Observatory in Antarctica reported some 28 energetic neutrinos that are almost certainly cosmic in origin.

How significant was the IceCube detection?

It marks the beginning of neutrino astronomy. Astronomy is not like other sciences; we usually don't get to put our quarry under the microscope or analyse it in the lab. We have to depend on feeble light from distant sources. By now, we've fairly well explored the electromagnetic spectrum. There are only two other cosmic messengers that we know of: gravitational waves and cosmic neutrinos.

What are the likely origins of the neutrinos Ice-Cube saw?

The two candidate sources are supermassive black holes at the hearts of galaxies and gamma-ray bursts, which are probably produced by the deaths of massive stars.

What else could cosmic neutrinos reveal?

There should have been neutrinos produced seconds after the Big Bang. With existing astronomy, we can look back only to about 380,000 years after the Big Bang. If we could detect these relic neutrinos, we could look back to within seconds of the birth of the universe. The problem is that they are now low in energy and therefore extremely difficult to detect. Present detectors are nowhere close to being sensitive enough to see them.

Can neutrinos capture the public imagination in the same way as the Higgs boson?

The Higgs has been a terrific story. But neutrinos allow us to probe some really profound questions and I think that makes them truly interesting. They're ready to take centre stage.

7
The lethal lightweight

The Higgs boson has a lower mass than expected. This fact could spell disaster for our entire universe.

The Higgs, but not as we know it

After the Higgs hoopla of 2012, particle physicists have been asking themselves whether that particle truly is the *pièce de résistance* of the standard model. And if it is, do we even want it?

As the standard model gradually took shape, it became clear how vital it was to find this particle. The model demanded that in the very early hot universe the electromagnetic and weak nuclear forces were one. It was only when the Higgs field emerged a billionth of a second or less after the Big Bang that the pair split, in a cataclysmic transition known as electroweak symmetry breaking. The W and Z bosons grew fat and retreated to subatomic confines; the photon, meanwhile, raced away mass-free and the electromagnetic force gained its current infinite range. At the same time, the fundamental particles that make up matter – things such as electrons and quarks, collectively known as fermions – interacted with the Higgs field and acquired their mass, too. An ordered universe with a set hierarchy of masses emerged from a madhouse of masslessness.

It's a nice story, but one that some find a little contrived. The problem is that the standard model is manifestly incomplete. The standard model's deficiencies suggest that what we need is not a standard Higgs at all, but something subtly or radically different – a key to a deeper theory.

Questions of identity

So far, the Higgs boson seems frustratingly plain and simple. The particle born on 4 July 2012 was discovered by sifting through the debris of trillions of collisions between protons within the mighty ATLAS and CMS detectors at CERN's Large Hadron Collider. For a start, it was spotted decaying into

W and Z bosons, exactly what you would expect from a particle bestowing them with mass.

Beyond that, a standard-model Higgs (see Figure 7.1) has to decay not just into force-transmitting bosons but also to matter-making fermions. Here, the waters are a little muddier. The particle was also seen decaying into two photons, which is indirect proof that it interacts with the heaviest sort of quark, the top quark: according to the theory, the Higgs cannot interact directly with photons because it has no electric charge, so it first splits into a pair of top quarks and antiquarks that in turn radiate photons.

It is too early to draw any firm conclusions. We know the new particle's mass fairly well (about 125 GeV, or 223 billionths of a billionth of a microgram), which pins down the rates at which it should decay into various particles to a precision of about 1 per cent, if it is the standard Higgs. Because of the limited number of decays seen so far, however, the measurement uncertainty on the new particle's decay rates is more like 20 per cent. The LHC will keep on working away for another 15 or 20 years to improve this.

For now, we are left with a particle that looks like the standard Higgs, although we cannot quite prove it. That leaves us facing an elephant in the accelerator tunnel: if it is the standard Higgs, how can it even be there?

The problem lies in the prediction of quantum field theory that particles spontaneously absorb and emit virtual particles by borrowing energy from the vacuum. Because the Higgs boson itself gathers mass from everything it touches, these processes should make its mass balloon from the region of 100 GeV to 10^{19} GeV. At this point, the Planck scale, the fundamental forces go berserk, and gravity – the comparative weakling of them all – becomes as strong as all the others. The consequence is a

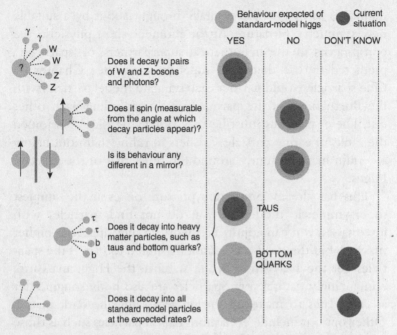

FIGURE 7.1 Steps to the Higgs: the standard-model Higgs boson has to pass many tests. Any deviation from its expected behaviour could be a sign of long-awaited new physics.

high-stress universe filled with black holes and oddly warped space-time.

Conspirators sought

One way to avert this disaster is to set the strength of virtual-particle fluctuations so they all cancel out, reining in the Higgs mass and making a universe more like the one we see. The only way to do that while retaining a semblance of theoretical dignity is to invoke a conspiracy brought about by a suitable new symmetry of nature. At the moment, most physicists see conspirators in the hypothetical superpartners, or 'sparticles', predicted by the theory of supersymmetry (see Chapter 8). One sparticle would partner each standard model particle, with the fluctuations of the partners neatly cancelling each other out. These sparticles must be very heavy: the LHC has joined the ranks of earlier particle smashers in ruling them out below a certain mass, currently around ten times that of the putative Higgs.

This has already put severe pressure on even the simplest supersymmetric models. If you do not find sparticles with low masses, you can adjust the theory so they appear at higher masses – but the goalposts cannot be shifted too far. If the sparticles get too heavy, they won't stabilize the Higgs mass in a convincingly natural way. Sparticles are also hotly sought after as candidates to make up the universe's missing dark matter. Other options include even more radical particles such as those inhabiting extra dimensions of space.

What if there is nothing but tumbleweed between the energies in which the standard model holds firm and those of the Planck scale, where quantum field theories and Einstein's gravity break down? How, then, would we explain the vast discrepancy between the Higgs's actual mass and that predicted by quantum theory? A lightweight hypothetical particle called the axion might prevent the Higgs mass from ballooning (see below).

Or should we just accept the Higgs mass is as it is? If things were different, the masses of all the particles and their interactions' strengths would be very different, matter as we know it would not exist, and we would not be here to worry about such questions. Such anthropic reasoning, which uses our existence to exclude certain properties of the universe that might have been possible, is often linked with the concept of a multi-verse – the idea that there are innumerable universes out there where all the other possible physics goes on.

To many physicists, this kind of argument is specious. But whether or not the Higgs mass is fixed by the fact of our existence, it could threaten our future, as we shall see.

On the brink of cosmic catastrophe

Our universe has been around for nearly 14 billion years, but it could vanish in the blink of an eye. If the fabric of space-time is in the precarious state that physicists call a false vacuum, it could collapse at any moment, taking us with it. The Higgs boson makes the universe stable – just. If the mass-giving particle were much lighter, the cosmos would quickly collapse in on itself.

The key to understanding how stable the vacuum is rests with the Higgs, boson and its associated field. Elementary particles get their masses by interacting with the Higgs field, and the mass of the Higgs boson depends on those particles as well. The heaviest of these, the top quark, has the biggest impact on the Higgs, and, based on recent measurements of both their masses, physicists can now use the properties of the Higgs field to deduce the state of the vacuum of space-time. The news is not great: our universe seems to be on the brink.

Like a ball rolling down a hill, the vacuum will eventually come to rest in the lowest possible energy state. If there is a small trough halfway down, however, the ball could get caught. The ball would have a kind of stability, but also the potential to roll further down. It seems we may be in just such a trough (see Figure 7.2) – although the measurements are not yet precise enough to say for sure.

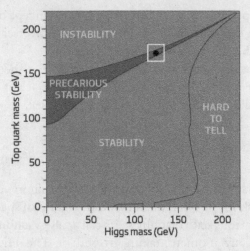

FIGURE 7.2 If the top quark were slightly heavier relative to the Higgs boson, the universe would have collapsed long ago.

The Higgs mass mystery

If we believe the standard model – our best theory of particle physics – a particle of dust should weigh as much as a well-fed elephant. That's because it predicts that Higgs should have a huge mass, which would in turn make other fundamental particles – electrons, quarks and neutrinos and so on – 100 quadrillion times heavier than we observe. This 'hierarchy problem' has haunted physicists for decades, and there has never been an easy answer. Our best models cannot explain why the mass-giving Higgs boson is so small.

Resurrecting a 40-year-old idea could put this riddle and two others to rest. The old front-runner, supersymmetry, has fallen from grace as the LHC keeps failing to find it, but the answer may have been under our noses all along. One hypothetical plaything of theorists since the 1970s could make the problem disappear.

Under the rules of quantum mechanics, the Higgs can temporarily shape-shift into all sorts of other particles, acquiring their masses in the process. Add up the effect of these quantum fluctuations and we get a value for the Higgs mass of about 10^{19} gigaelectronvolts – roughly the same as an eyelash. We knew it could not be so high, because in turn it would make other particles super-heavy. And when researchers at the LHC found the Higgs at a paltry 125 GeV, it confirmed that there is something massive missing from our understanding of the Higgs.

A blunt resolution to this huge discrepancy is to assume that it has some sort of inherent mass, independent of all the quantum fluctuations, and that this is also very large. Then all the separate contributions to the overall mass could almost cancel out, with the remainder being the mass we actually measure.

A coincidence of cosmic proportions

What are the chances of two unrelated, huge figures cancelling almost perfectly? Looked at this way, the hierarchy problem is a coincidence of cosmic proportions. It's a bit like driving a golf ball towards the green, only to see it deflect off a tree into a bunker, bounce into a lake and then rebound to land just beside the club you hit it with. And the improbability is not just limited to the boson. Because the masses of all the fundamental particles scale with the Higgs, the problem affects just about everything.

Theorists loved supersymmetry because it appeared to offer a solution. But what happens if the sparticles it predicts continue to be a no-show?

In 2014 Kaplan's colleagues Surjeet Rajendran at the University of California, Berkeley and Peter Graham at Stanford University devised a new type of experiment to detect an alternative: the axion. If they exist, axions are light, electrically neutral particles that generate their own unique force. They were first proposed more than 40 years ago, to explain a vexing issue known as the strong charge parity problem, which is about why the strong force maintains CP symmetry when the weak does not. They are also prime candidates for dark matter, the invisible stuff that accounts for 80 per cent of the matter in the universe.

No axion detector yet devised has seen anything. Rajendran and Graham had a new design up their sleeves and were about to apply for funding to build it when a team claimed to have detected long-sought evidence of a period of rapid inflation early in the universe, using the BICEP telescope near the South Pole. That seemed to rule out the existence of axions, but Rajendran was less sure. He eventually worked out how to square the BICEP result with the existence of axions. It required the axion to have had a large mass in the early universe, which then petered out.

The BICEP claim turned out to be wrong, but the episode had got Rajendran thinking. If the axion mass could decrease over time, could that apply to other particles such as the Higgs boson, too? Rajendran, Graham and Kaplan applied this idea to the hierarchy problem by suggesting that the masses of the axion and Higgs are linked, like two wheels on an axle. Both particles could start off in the early universe at the gigantic mass predicted by the standard model, and then slowly roll downhill.

Why would the Higgs mass grind to a halt at 125 GeV rather than keep falling? The team arrived at the idea that the mass-giving nature of the Higgs itself may have switched on only when its mass dipped below a certain value. When that happens it stays locked in place. The Higgs suddenly gives mass to quarks, which interact with axions via the strong force. That puts a cap on the axion mass, and in turn on the Higgs.

The team suggested tweaking the axion's name to 'relaxion', given how it could cause the mass of the Higgs to relax nicely down to its observed value. Although it may seem contrived, the relaxion idea does have one advantage over other answers to the hierarchy problem. There are detectors out there hunting for axions now, such as the recently revamped axion dark matter experiment at the University of Washington, Seattle.

And Graham, Kaplan and Rajendran are working on an audacious new idea. Could relaxion theory solve a fourth big mystery? The cosmological constant is the number physicists plug into equations to represent the dark energy pushing space apart, inflating the universe. A very small number is needed for the expansion to tally with observations, and yet quantum field theory implies that it should be gargantuan. Maybe a similar sort of relaxation argument might answer the problem.

At the Higgs Nobel bash

In December 2013 New Scientist's *Valerie Jamieson was invited to join the ceremony when Peter Higgs and François Englert were jointly awarded a Nobel Prize in Physics. Here she describes the experience.*

'Scientists dream of a call from Stockholm. Mine came when I was vacuuming the living room. "Would you like to come to the banquet?" asked a man from the Nobel Foundation. My heart soared. I hadn't won a Nobel Prize, but this is one award ceremony that anyone curious about the universe would enjoy, for the discovery of how the building blocks of matter get their mass.

'For me, it was also personal. Long before the Higgs had been detected, I walked inside the tunnel that now houses the LHC, treading the ground where protons would soon be smashed together at the speed of light. I felt knee-high to a grasshopper when I looked up at the detectors ATLAS and CMS, each several storeys high, that would go on to ensnare the Higgs. When the first protons set off in 2008, I was back again. Attending the Nobel ceremony was the next chapter.

'As I arrived at the Stockholm concert hall on 10 December, I realized that this was Sweden's Oscars – complete with crowds lining the square outside and two nude protesters at the entrance. Inside was like a Who's Who of physics and a living history of the standard model, our best theory of the universe's particles and forces.

'There was Carlo Rubbia, who won the 1984 prize for discovering the W and Z particles, which carry the weak force responsible for radioactive decay. I also picked out Gerhard 't Hooft, who won in 1999 for taming the standard model's unwieldy equations, and David Gross and Frank Wilczek, who won in 2004 for their work on the strong force, which holds together atomic nuclei.

'These minds built the standard model – but it was incomplete until the Higgs was discovered. Fittingly, I had to wait until the very end of the evening to meet Peter Higgs. He and Englert were leading a fight against the dress code: "They wanted us to wear these ridiculous shoes. François and I started a rebellion," said Higgs. "I thought we might be thrown out."'

8

Superparticles and beyond

With the standard model seemingly tied up, what's next? Physicists are anxiously scanning the horizon for new and monstrous particles.

What the LHC is really looking for: SUSY

Finding the Higgs in 2012 was the capstone of an edifice that particle physicists have been building for half a century – the standard model. But since then, as the LHC has kept running and increasing its energy, particle physicists have been getting more and more anxious at the lack of new discoveries. We know that the standard model is incomplete. It has nothing to say about the fourth fundamental force of nature, gravity; it is silent on the nature of dark matter; it needs fine-tuning to bring the mass of the Higgs down to a manageable level (see Chapter 7). So the model must be just part of something much bigger. If the LHC finds nothing beyond the Higgs, particle physics will be at a dead end, with no clues where to turn next.

Most of all, physicists were hoping to find evidence of the overarching theory known as supersymmetry, or SUSY, to set them on the path to unification. In today's universe, the three forces in the standard model have very different strengths and ranges, but in the 1960s Steven Weinberg, then at Harvard University, showed with Abdus Salam and Sheldon Glashow that, at the high energies prevalent in the early universe, the weak and electromagnetic forces have the same strength; in fact, they unify into one force. The expectation was that, if you extrapolated back far enough towards the Big Bang, the strong force would also succumb and be unified with the electromagnetic and weak force in one single superforce (see Figure 8.1). The standard model can make that happen – but only approximately. This not-so-exact reunification soon began to bug physicists.

Then supersymmetry made its appearance. It debuted in the work of Soviet physicists Yuri Golfand and Evgeny Likhtman; and Julius Wess of Karlsruhe University in Germany and Bruno

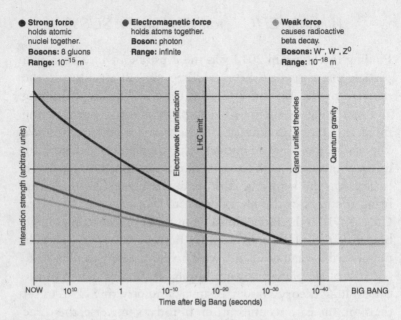

FIGURE 8.1 The forces we know today have very different strengths. But if we could roll back time to the Big Bang, or simulate its conditions inside a particle accelerator, we would see them becoming similar in strength and eventually become one superforce.

Zumino of the University of California, Berkeley, brought it to wider attention a few years later.

The aim is to extend physicists' favourite simplifying principle, symmetry, and show that the division of the particle domain into fermions and bosons is the result of a lost symmetry that existed in the early universe. Today, each fermion would be paired with a more massive supersymmetric boson, and each boson with a fermionic super-sibling (see Figure 8.2). For example, the electron has the selectron (a boson) as its supersymmetric partner, while the photon is partnered with the photino (a fermion).

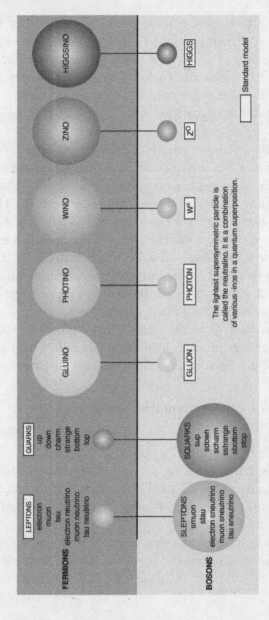

FIGURE 8.2 The particle zoo of supersymmetry, which doubles the number of particles, giving each fermion a massive boson as a superpartner and vice versa

The key to the theory is that in the high-energy soup of the early universe, particles and their superpartners were indistinguishable. Each pair co-existed as single massless entities. As the universe expanded and cooled, this supersymmetry broke down. Partners and superpartners went their separate ways, becoming individual particles with a distinctive mass all of their own.

Supersymmetry solves what is known as the hierarchy problem (see Chapter 7), where, according to quantum field theory, the Higgs mass and the mass of other particles should be vast. SUSY can tame all the pesky contributions from the Higgs's interactions with elementary particles, the ones that cause its mass to run out of control. They are simply cancelled out by contributions from their supersymmetric partners. And with supersymmetry added to the mix, the curves representing the strengths of all three forces can be made to come together in the early universe with stunning accuracy.

This was enough to convert many physicists into true believers. But things became really interesting when they began studying some of the questions raised by the new theory.

The hunt for supersymmetric particles

One pressing question was the present-day whereabouts of supersymmetric particles. Electrons, photons and the like are all around us, but there is no sign of selectrons and photinos, either in nature or in accelerator experiments. If such particles exist, they must be extremely massive, requiring huge amounts of energy to fabricate.

Such huge particles would have decayed into a residue of the lightest, stable supersymmetric particles, called neutralinos. The neutralino has no electric charge and interacts with normal matter timorously by means of the weak nuclear force. No surprise then that it has eluded detection so far.

When physicists calculated exactly how much of the neutralino residue there should be, they were taken aback. It was a huge amount – far more than all the normal matter in the universe. It seemed that neutralinos fulfilled all the requirements for the dark matter that astronomical observations persuade us must dominate the cosmos.

All this seems to point to some fundamental truth locked up within the theory. But mathematical beauty and promise are not enough: you also need experimental evidence. Circumstantial evidence for supersymmetry might be found in various experiments designed to find and characterize dark matter passing through the Earth. These include the SuperCDMS experiment, in a mine near Sudbury in Canada, and the XENON1T experiment beneath the Gran Sasso Mountain in central Italy. Space probes like NASA's Fermi satellite are also scouring the Milky Way for the telltale signs expected to be produced when two neutralinos meet and annihilate.

The best evidence would come if we could produce neutralinos in an accelerator. Because neutralinos barely interact with other particles, they will evade the detectors, but that may make them relatively easy to find as the energy and momentum they carry will appear to be missing.

We are not entirely sure how muscular that accelerator would need to be. The mass of the superpartner depends on precisely when supersymmetry broke apart as the universe cooled. But the simplest, most aesthetically pleasing forms of SUSY (known as constrained minimal) predict that quark superpartners (squarks) have masses below 1 TeV. That is squarely in the LHC's energy range – but it has not seen any. What has gone wrong?

Vanilla SUSY

Searches for supersymmetric particles at the LHC so far have concentrated on finding the 'stop', the squark equivalent of the standard model's top quark. In most versions of SUSY, the stop is the lightest squark. In 'plain vanilla' constrained minimal SUSY, the other squarks are not much heavier. They should all be produced at the LHC, with the heavier ones decaying into the stop, resulting in a stop deluge that would be hard to overlook – exactly what has not been seen.

Perhaps we need to turn to more complex variants of SUSY that need more assumptions and free parameters to make them work. Some of these deliver higher values of squark and gluino masses in regions the LHC has not yet tested, while still giving a stop mass that is below 1 TeV. If these models are right, the LHC would have produced far fewer heavier squarks, or perhaps none at all. Then we just need to refine the search – looking, for example, for the stop being produced directly rather than as a decay product of something heavier.

Suddenly, the talk is of fine-tuning a welter of free parameters in supersymmetric models to get the right results – just the sort of fudge SUSY was designed to avoid. That invites a more radical conclusion: perhaps nothing can save SUSY?

Some physicists have been dusting off old models that might replace it (see Figure 8.3). There's the axion theory discussed in the previous chapter. Then there is the brainchild of Steven Weinberg, now at the University of Texas, Austin. In 1979 he was working with Leonard Susskind of Stanford University, and their starting point was the humdrum proton. The proton is made of quarks bound together by gluons, which mediate the strong nuclear force, yet most of its mass comes not from the quarks but from the energy contained in the bonds between them (see Chapter 4). These colour interactions are

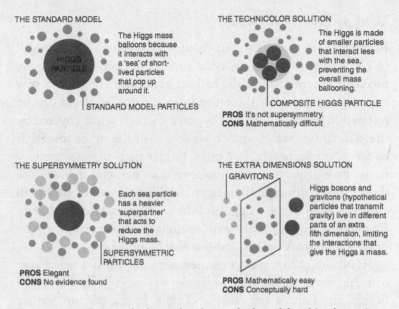

THE STANDARD MODEL

HIGGS PARTICLE

The Higgs mass balloons because it interacts with a 'sea' of short-lived particles that pop up around it.

STANDARD MODEL PARTICLES

THE TECHNICOLOR SOLUTION

The Higgs is made of smaller particles that interact less with the sea, preventing the overall mass ballooning.

COMPOSITE HIGGS PARTICLE

PROS It's not supersymmetry.
CONS Mathematically difficult

THE SUPERSYMMETRY SOLUTION

Each sea particle has a heavier 'superpartner' that acts to reduce the Higgs mass.

SUPERSYMMETRIC PARTICLES

PROS Elegant
CONS No evidence found

THE EXTRA DIMENSIONS SOLUTION

GRAVITONS

Higgs bosons and gravitons (hypothetical particles that transmit gravity) live in different parts of an extra fifth dimension, limiting the interactions that give the Higgs a mass.

PROS Mathematically easy
CONS Conceptually hard

FIGURE 8.3 Sorting the hierarchy: the standard model and its alternatives

the expression of the strong nuclear force at the low energies of today's universe. If a similar mechanism had been at work at much higher energies in the early universe, Weinberg and Susskind reasoned, that could explain why elementary particles such as quarks themselves have mass, without needing the Higgs. It was a bright new prospect they named technicolor.

But technicolor's mathematics was tough, and the few testable predictions that could be extracted from it did not tally well with experimental results from the Large Electron–Positron Collider (LEP), CERN's principal accelerator until 2001. Tweaks to the theory allayed some of those problems, but technicolor's lustre soon faded.

Why is gravity so weak?

In the late 1990s Raman Sundrum, together with Lisa Randall of Harvard University, suggested another option. The hierarchy problem has to do with the ballooning of the Higgs mass way beyond the masses of the other known particles, but it can be restated in another way: why is gravity, which is not covered by the standard model, so much weaker than the other forces? It is, for example, nearly 10^{34} times punier than the electromagnetic force. If gravity were stronger, then particles that acquire their masses through the Higgs mechanism would be much closer in mass to the Higgs, and the problem would melt away. Conversely, find a theory with an inbuilt explanation for why gravity is as weak as it is, and the problem dissolves.

Randall and Sundrum's mathematics suggested a novel way to bring about the desired weakness: an unseen fifth dimension besides the four of our space and time. In this picture, we are like ants living on the two-dimensional surface of a piece of paper. They scuttle around, unaware that their world also has an infinitesimally small third dimension, the paper's thickness. Randall–Sundrum models suggested that the particles mediating gravity, gravitons, prefer to populate one side of a 5D universe – one side of a sheet of paper, if you will. Higgs bosons, meanwhile, hang out on our side. This limits the interaction of gravitons with particles such as electrons and quarks that get their mass through the Higgs mechanism, and so gravity appears weak in our 4D approximation of space-time. In a full 5D view, meanwhile, it is just as strong as all the rest of the forces.

One suggestion is that we should mix technicolor and Randall–Sundrum models together. In 1997 Juan Maldacena of the Institute for Advanced Study in Princeton, New Jersey, showed how an intractable theory of strong interactions in a four-dimensional space-time such as our own can be made a lot more tractable by adding an extra dimension. Randall and Sundrum saw that this could form a bridge between their theory and technicolor, and since then they have been working out the details.

The most promising technicolor-like theory does not get rid of the Higgs entirely, but says it is not an elementary particle. Instead, it is a composite of other, new elementary particles, a 'bound state' rather as a proton is really a bound bunch of quarks and gluons. It makes predictions about how the composite Higgs decays, which the LHC should be able to test. It also predicts new particles: resonances of the known particles with masses greater than 1 TeV, which have extra mass-energy from their vibrations in the fifth dimension.

The hope is that CERN's behemoth might yet find something – SUSY's stop quark, Randall and Sundrum's resonances or something completely different – that points the way towards a newer and bigger theory of matter.

Shadow worlds: our first glimpse of dark forces?

Spare a thought for the dark-matter hunters. Judging by cosmological observations of the way galaxies and clusters of galaxies whirl around, this elusive stuff makes up the bulk of the universe's matter. But every time physicists seem on the verge of trapping it, it slips away. They don't see the expected signals, or they spot something exciting only to watch it fade into background noise. Each time it's the same: put on a brave face, go back to the drawing board and begin the hunt again.

Here it might be time for a change of tack. Instead of going after a single species of dark-matter particle, maybe we should be looking for a menagerie of dark particles and forces – a whole new 'dark sector'. After all, there is no reason to think dark matter will be any less intricate than the visible stuff we consider ordinary, with its panoply of particles from electrons to quarks.

The talk is of an entire shadow world in which invisible particles influence one another through forces unfelt by the familiar stuff of stars and planets and us. And if the faint hints of a dark force emerging from one lab stand up to scrutiny, this shadow realm may already have revealed itself.

From as early as the 1930s, astronomers could see that galaxies orbit each other much faster than expected, given the gravitational tug produced by their visible stars. Forty years on, we spotted that stars within galaxies also seemed to rotate too fast. And in simulations of the universe, the gravity of normal matter is not strong enough to pull galaxies and clusters together out of primordial gas. Either the laws of gravity drawn up by Newton and Einstein required a substantial rewrite, or some invisible form of matter was producing more gravitational heft. Most astronomers favour the second option – dark matter. They calculate that dark matter outweighs normal, visible matter by about four to one.

Whatever this stuff is made of, it must have mass so that it feels and generates gravity; but no electric charge, so it does not interact with light. For decades, the leading candidate has been the weakly interacting massive particle, or WIMP, usually thought to be something much heavier than a proton. This could be supersymmetry's neutralino, a stable particle that might have been created in huge numbers in the first moments of the Big Bang.

Searching for WIMPs

There may be some evidence for these particles in the gamma-ray glow of a nearby dwarf galaxy (see 'A hint of WIMPs?' below), but here on Earth a long line of exceedingly sensitive detectors has failed to record a single WIMP. For some researchers, it's time to move on.

There are other reasons to consider more extravagant alternatives. We can now measure the rotation of galaxies in enough detail to work out how dark matter is distributed within them. Simple WIMP models suggest that it should be very dense in the middle of galaxies. The latest observations, however, show that it is spread more evenly. One way to explain that is by appealing to a force that acts only between dark-matter particles, pushing them apart.

That could be a game changer. Once you start thinking about forces acting just between dark-matter particles, you are led into a whole new arena, says Jonathan Feng at the University of California, Irvine. You can think about a zoo of dark particles and forces all of its own.

The idea of a dark sector is not entirely novel. In 2006 astronomers studying the Bullet Cluster, an ongoing smash-up between two groups of galaxies (see Figure 8.4), proposed that the collision speed was too high for the gravity of the matter

FIGURE 8.4 Galactic smash-ups like this one, the Bullet Cluster, could reveal whether dark matter is more complex than we thought.

involved – dark and ordinary – to be solely responsible. They thought that the additional pull must be coming from a force of attraction between dark-matter particles.

A hint of WIMPs?

Dark matter might not be as gloomy as its name suggests. If this mysterious substance is made of weakly interacting massive particles (WIMPs), as most physicists believe, then they would come in matter and antimatter versions. When the two come into contact, they would produce a shower of high-energy photons known as gamma rays.

Galaxies tend to be crowded with billions of stars, making it almost impossible to rule out other sources for the gamma rays, but in the past few years astronomers have discovered a nearby population of ultra-faint dwarf galaxies with no more than a few hundred million stars. These mini-galaxies are also thought to hold unusually high concentrations of dark matter, making them the ideal place to look for its gamma-ray glow.

In 2015 Alex Geringer-Sameth at Carnegie Mellon University in Pittsburgh, Pennsylvania, and his colleagues looked at a newly discovered dwarf galaxy called Reticulum II, just 100,000 light years away. In archived observations from NASA's Fermi gamma-ray space telescope, they found what appeared to be an excess of gamma rays.

Critics say there could be hidden gamma-ray sources beyond Reticulum II. There are no plans for new gamma-ray telescopes to provide more accurate observations, and unless we discover more nearby dwarf galaxies to check in Fermi's archive, this remains just a tantalizing suggestion of WIMPs in space.

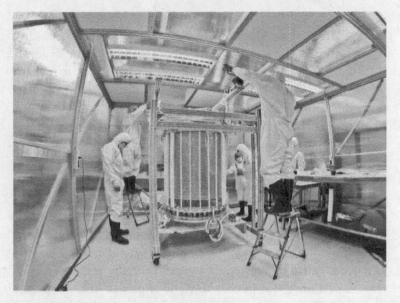

FIGURE 8.5 Even the cleanest, most sensitive detectors have failed to lay a glove on dark-matter particles.

A dark-force carrying particle

More detailed simulations proved that the speed of the Bullet Cluster collision was not beyond what we might expect. But the suspicion of dark forces never went away, with researchers suggesting that anomalies thrown up by particle experiments on Earth might also hint at their existence. For example, a long-standing discrepancy between theory and experiment over the magnetic properties of ordinary matter particles called muons, a heavier version of the electron, might be explained by invoking a dark-force-carrying particle. Now Feng thinks we might have found the most compelling evidence for such a particle so far, in a nuclear physics lab in Hungary.

Attila Krasznahorkay of the Institute for Nuclear Research at the Hungarian Academy of Science in Debrecen leads a team looking at the radioactive decay of beryllium-8 nuclei. Beryllium is a naturally occurring light element that is stable when its nucleus contains four protons and five neutrons. But with just four of each, the isotope Be-8 splits into two helium nuclei in the blink of an eye. Previous experiments had hinted at something odd about this particular decay, and Krasznahorkay and his colleagues wanted to pin it down.

To make Be-8, they fired protons at a wafer-thin sheet of lithium-7. The beryllium decayed, releasing pairs of electrons and their antimatter counterparts, positrons. In standard particle theory, most of those pairs should be emitted in roughly the same direction as the incoming proton beam. But the Hungarians found that there were two unexpectedly prominent side streams, coming out almost at right angles to their expected direction. This was the sort of behaviour you would expect if the decay created a slow-moving particle that lived for short time before itself decaying into an electron and positron, which it would spit out in almost opposite directions.

When the team calculated the mass of this hypothetical particle, they found that it fitted nothing in the standard model of particle physics. Instead, their numbers suggest that it has a mass of around 17 MeV – just 33 times that of an electron and far lighter than any WIMP. No known force of nature could create such a particle.

Dark photons

Having investigated the anomaly for three years, the team published their results in 2015. They refer to their particle as a dark photon. By analogy with the way the photon carries electromagnetism, this particle would carry an unknown force between dark-matter particles.

Taking the results at face value, Feng and his colleagues sought their own explanation. They also wanted to address a nagging doubt: given that the Hungarian team spotted this putative new particle with an experiment well within the capabilities of most physics labs around the world, why had no one else noticed anything before?

The hypothetical dark photon, as well as carrying the dark force between dark-matter particles, should also carry a little bit of ordinary electromagnetism. It should therefore occasionally interact with the protons and electrons in normal matter. But when Feng and his colleagues calculated the strength of this interaction, things became more complex. 'There was no way this could be a dark photon,' says Feng. 'If it were, we should have seen hundreds and thousands of other effects in other experiments and particle accelerators.'

If not a dark photon, what was it? Feng's team searched for other ways a dark particle could be interacting, albeit slightly, with familiar matter to cause the anomalous beryllium decay. They found that, to be consistent with everything we have seen in experiments designed to characterize the known forces of nature, it must interact not with protons and electrons, as a conventional photon does, but with the neutrons inside the beryllium nuclei. This is a property beyond the scope of physics as we know it, which might explain how the particle slipped by unseen in previous dark matter searches. Feng's team call the interloper a protophobic X boson.

Not everyone is convinced of claims of a whole shadow world beyond the visible material universe, but Feng's theory can be tested. The DarkLight experiment at the Thomas Jefferson National Accelerator Facility in Newport News, Virginia, is already searching for particles in the mass regions where Feng's team calculated it should be. The LHCb experiment at CERN will also look for it in the decays of quarks and their antimatter counterparts.

Six ways to make dark matter

1 **WIMPs** The textbook solution to dark matter is that it is a soup of slow-moving heavy particles, WIMPs. They can certainly help explain the way galaxies rotate and the creation of galaxies and clusters in the first place. But no detector has yet definitely found a WIMP, and the hints they have seen indicate WIMPs ten times lighter than expected. If such lightweight WIMPs really do exist, they sit at the extreme low end of the possible range.

2 **MACHOs** This is the idea that dark matter is just normal stuff hiding at the edges of galaxies – 'massive astrophysical compact halo objects' that are so dim as to be invisible. Candidates include black holes or failed stars. Alas, MACHOs could account for only a tiny fraction of the universe's missing mass.

3 **Macros** It could be that the dark stuff is made of dense clumps of quarks, the particles that, in pairs or triplets, form ordinary matter. These 'macros' could be as dense as neutron stars and extremely heavy. We might be able to spot them one day by deploying seismometers on the Moon.

4 **Axions** A punier version of the WIMP, axions would interact even less with ordinary matter. While projects such as the Axion Dark Matter Experiment extend the search, the jury is still out.

5 **Sterile neutrinos** Neutrinos pass through other matter almost as if it doesn't exist, but they are too light and zippy to be dark matter. Sterile neutrinos would be a heavier, more aloof version.

6 **Gravitinos** In attempts to meld general relativity with the theory of supersymmetry, the graviton mediates the force of gravity and the gravitino is its hypothetical superpartner. It nicely fits the bill for a dark-matter particle. The trouble is that there is still no sign of the many heavy partner particles predicted by supersymmetry.

The oddball particles to solve physics puzzles

The standard model leaves many questions unanswered. Why does matter dominate antimatter in our cosmos? What is the true nature of gravity? What is dark matter? Attempts to answer such questions lead physicists time and time again to the same expedient: invent a new particle ...

Leptoquarks

In 1994 a team of physicists were colliding electrons head on with protons at the DESY laboratory in Hamburg, Germany, when they saw an electron apparently turn into its heavier counterpart, a muon. Such a transformation would require the electron to take energy to make mass from the proton – something unheard of in the standard model, where electrons and protons are very different sorts of particle. Protons are composites bound by the strong force; electrons and muons are elementary particles, collectively known as leptons, that do not feel the strong force at all.

One possibility is that the collisions created a heavyweight crossbreed known as a leptoquark. In some grand unified theories, which roll three of the four forces of nature into one, when an electron hits a proton these leptoquarks can form and decay to a muon and a quark.

There were no further sightings, and the excitement faded. Yet the lure of grand unified theories remains, and the search for leptoquarks continues at the LHC today.

Stringballs

String theory is a popular shot at bringing together two disparate scales – the tiny world of quantum particles, where the standard model holds sway, and the cosmic distances over which gravity acts (see Chapter 9). It holds that particles such as electrons and quarks are really strings of energy a mere 10^{-35} metres long vibrating in different ways. If so, the LHC or a future accelerator might create stringballs. These are made when two strings slam into one another and, rather than combining to form a stretched string, make a tangled ball. Finding them would prove string theory, which requires extra dimensions of space in addition to the three we know about.

Inflatons

Why is space so smooth and why are the contents of the cosmos so evenly distributed? The standard explanation is that, just after its birth, the universe went through a period of breakneck expansion in which regions of space were pulled apart faster than the speed of light, ironing out all the wrinkles. The driving force behind this inflation was a hugely energetic field that briefly dominated the cosmos before dissolving into other matter and radiation. Quantum theory says that every field has an associated particle – in this case the inflaton. Its existence would have some intriguing implications. Quantum fluctuations in the inflaton field make it very difficult to turn off completely, so parts of the original cosmos will still be inflating, making for a multiverse of independently developing universes.

Direct evidence for the inflaton won't be coming any time soon, because they must have enormous mass. You would need an accelerator producing at least a trillion times the LHC's energy.

Wimpzillas

Physicist Rocky Kolb was doing his grocery shopping in Warrenville, Illinois, one day, and wondering what he should call the dark-matter particle he and his colleagues had just invented. A movie poster on a passing bus provided the answer. It was 1998 and the *Godzilla* remake had just been released. The wimpzilla was born.

Within the universe's first second, during the period of inflation, the expansion of space could have ripped particles out of the vacuum. Kolb and his colleagues calculated that among them might have been dark particles weighing a billion times more than a standard WIMP, with masses of many exa-electron volts (10^{18} eV).

Their monstrous mass means that wimpzillas would be exceedingly rare. Like inflatons, they could not be made in particle accelerators and are unlikely to amble into one of the underground detectors looking for WIMPs. But they might leave subtle footprints in the cosmic microwave background radiation, the Big Bang's afterglow that suffuses the sky.

Pole alone: the quest for a north without a south

North and south magnetic poles exist, just as positive and negative electric charges do. But from the humblest bar magnet to the Earth's mighty interior dynamo, magnetic poles only ever crop up tied together in pairs. Chop a magnet in half and, like Walt Disney's sorcerer's apprentice with his magic broom, you forge two new complete magnets, each with a north and a south pole.

Although nobody has ever seen a north pole without its accompanying south, or a south without its north, many theorists remain hopeful of finding such a monopole. For one thing, it would complete the equations collated by James Clerk Maxwell in the 1860s. These encapsulate the idea that electricity and magnetism are two manifestations of the same thing: the fundamental force of electromagnetism. Maxwell's equations allow individual, freely moving electric charges, which nature supplies in abundance in the form of particles such as electrons and protons. Similar freewheeling magnetic charges would give an aesthetically pleasing symmetry in the equations – but, faced with the lack of any sightings, Maxwell eschewed beauty and wrote the freely moving monopole out of his equations.

Monopoles made a comeback thanks to Paul Dirac, the notoriously word-shy British theorist who was obsessed with mathematical beauty. In 1931, applying the ideas of quantum theory to Maxwell's classical electromagnetism, Dirac showed that, even if there was just one magnetic monopole in the entire universe, its existence would explain why all the electric charge we see comes in the same bite-sized chunks of +1 or −1.

Forty years on, physicists were seeking ways to unify electroweak interactions with the strong nuclear force. Gerard 't

Hooft and Alexander Polyakov independently showed that monopoles were essential, or else such a grand unified theory would allow particles to have all sorts of electrical charges.

The monopole quest

Researchers have looked for monopoles everywhere, from Antarctic ice to Moon rocks. But the closest we've come to the real thing was on Valentine's night in 1982. That was when physicist Blas Cabrera saw a promising event in a monopole detector he had set up in a basement at Stanford University in California. It proved to be a one-night stand, prompting some wag to post Cabrera this loving note exactly a year later: 'Roses are red/Violets are blue/The time has now come/For monopole two.'

These days, Cabrera's monopole and a couple of other even more ambiguous sightings are regarded as experimental blips. But Dirac's calculations provide a ready-made excuse for the monopole's absence. They show that the smaller the unit of electric charge is, the larger the unit of magnetic charge must be. Because the basic electric charge is so small, the basic magnetic charge is so large that it would take an implausible amount of energy to make a particle carrying it.

According to the standard model, the only time enough energy was around to make monopoles in any abundance was in the first infinitesimal fraction of a second after the Big Bang. At about the same time the period of rip-roaring cosmic expansion known as inflation is thought to have occurred, which would have scattered monopoles to the four winds. There could be just one in the entire volume of the visible universe, says theorist Joseph Polchinski at Caltech.

But others hope that they remain a little more common. At the LHC, Jim Pinfold of the University of Alberta is looking for them with an experiment called MoEDAL (short for the Monopole and Exotics Detector at the LHC). Its main component is a series of metal cases attached to the walls around the platform. Neatly outlined in yellow masking tape, each trails a curl of wires to the floor.

Within each of the metal boxes are detectors consisting of a series of stacked layers of plastic that act as a form of photographic plate. Because a monopole must carry a huge magnetic charge, it would rip through the polymer bonds of the plastic, etching a trail whose size, shape and alignment would reveal the particle's character.

Even though the mini fireballs produced at the LHC are the most energetic ever made in a particle accelerator, they fall far short of what is needed to make a monopole if you take the standard model at face value. But Pinfold's experiment is no fool's errand. Many attempts to improve on the standard model predict a much lighter monopole.

Monopoles should be stable, so Pinfold's experiment includes trapping detectors in which any passing monopole might be bottled and kept for further tests – true exotics in the particle zoo.

Where can we look for monopoles?

The hunt for magnetic monopoles has been far-reaching. Here's where we have looked and failed to find them:

Trapped in ...

- Moon rocks from the Apollo 11 mission
- Antarctic meteorites
- volcanic rocks

Passing through ...

- experiments looking for high-energy neutrinos and cosmic rays
- dedicated monopole detectors
- cosmic microwave background radiation

Made in ...

- particle colliders
- magnetic materials

9
Pieces of gravity

Gravity is a glaring omission in our current best model of the fundamental forces of nature. There are some promising avenues to explore, but how close are we to a theory of everything?

Still no theory of everything

The standard model of particle physics describes most of the fundamental forces of nature in terms of quantum field theory: flitting virtual bosons carry the forces. But it does not encompass gravity. Instead, we use Einstein's theory of general relativity to describe gravity in completely different terms, as a consequence of curved space-time.

A quantum theory of gravity seems necessary to fully understand black holes and the origin of the universe, where gravity becomes so powerful that general relativity breaks down, giving infinity for every answer. Such a theory should give us a much deeper insight into the nature of space and time.

However, when you sit down and try to describe gravity with a quantum field theory, you soon have a big problem. Calculating any quantum particle process is enormously complex, as you must add up the infinite variety of ways virtual particles can be produced. Sometimes the sum of all these processes is finite; but at other times it blows up and you get infinity. For example, the quantum theory of beta decay gave infinities – until physicists developed electroweak theory. This tamed the mathematics by adding a quantity of undiscovered massive particles, the W, Z and Higgs bosons, which cancelled out the infinities.

This success led physicists to believe that this strategy was something like a general prescription for developing quantum theories: if your model produces infinities, then you can just add new, more massive particles to solve the problem. Suppose gravity is made of quantum particles called gravitons, as light is made of photons. As we sum over all possible histories, the calculations rapidly spiral as expected into a chaos of infinities (see Figure 9.1).

This time, cancelling the infinities requires inventing a new particle with a mass 10 billion billion times that of a proton.

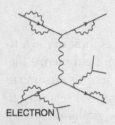
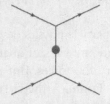

ELECTRON

GRAVITON

Particles such as electrons can interact by producing and exchanging massless photons in countless ways, often resulting in infinities in the calculations.

The situation is saved by the existence of heavier particles – the **W** – **Z and Higgs bosuns,** which cancel out the infinities.

Performing the same trick with graviton interactions re- quires a particle so massive it acts like a **black hole** – and all calculations are off again.

FIGURE 9.1 Gravitons are conjectured quantum particles of gravity – but theories incorporating them tend to be unruly.

As with all virtual particles, they must pay this energy back by disappearing again – and the more they borrow, the more quickly that must happen, so these particles are very short-lived. They cannot get very far, and so occupy only a minute amount of space. So much mass in such a small space forms a black hole – containing a singularity of infinite density.

Attempts to get round this fundamental roadblock have led us to destinations such as string theory, which assumes that all particles are manifestations of more fundamental vibrating strings, and loop quantum gravity, which suggests that space-time itself is chopped up into discrete blocks.

String sounds

In conventional quantum field theories, the universe is made from fundamental point particles devoid of size, shape or structure. In superstring theory, the building blocks of matter are instead one-dimensional strings that live in a universe with ten dimensions. These extra dimensions are not like the other four, but instead curled up into a circle too small to see. In fact, at around about 10^{-35} metres across, they are too small to probe even with the most powerful particle accelerators today.

Just like violin strings, superstrings can vibrate in various modes. We see each mode of vibration as a different elementary particle. String theory is one way to unify the strong force with electromagnetism and the weak force. But it goes much further. String theory works as a quantum theory of gravity because string vibrations can describe gravitons, the hypothetical carriers of the gravitational force. Even better, the interactions between strings are in some sense smoother than between point-like particles, and so string theory does away with the infinities and anomalies that plagued previous attempts to apply quantum field theory to general relativity.

Thanks to work by Ed Witten in the 1990s, string theory has now been incorporated into M-theory. This is an overarching mathematical framework that lives in 11-dimensional spacetime, involving higher-dimensional extended objects called p-branes, of which strings are just a special case.

The precise way those extra dimensions are curled up dictates the appearance of our four-dimensional world: including how many generations of quarks and leptons there are, which forces exist, and the masses of the elementary particles. A puzzling feature of M-theory is that there are many ways to curl up these dimensions, leading to a plethora of possible universes. So there could be many universes out there with different laws of physics, and one of them just happens to be the one we are living in.

M-theory remains unproved, of course, but it has had some successes. For example, in 1974 Stephen Hawking showed that black holes can radiate energy due to quantum effects, meaning that they have temperature and another thermodynamic property called entropy, which is a measure of how disorganized a system is. Hawking showed that a black hole's entropy depends on its area. It should be possible to work out its entropy by accounting for all the quantum states of the particles making up a black hole, but all attempts to describe a black hole in this way had failed – until M-theory came along, exactly reproducing Hawking's entropy formula.

Is the lack of superpartners seen at the LHC as bad news for superstring theory?

The 'super' in superstring does indeed refer to supersymmetry, which is a crucial element of the theory. The fact that the LHC has so far failed to detect any superpartners is often said to cast doubt on supersymmetry, but in fact such light superpartners are only necessary if SUSY is the solution to the hierarchy problem (see Chapter 8). String/M-theory is compatible with supersymmetry becoming evident only at much higher energies.

When loops become strings

Two of our most promising theories of quantum gravity have been competing for decades. What if they are actually the same thing?

String theory may be successful, but it has its share of hangups. Its extra dimensions can be folded up in so many ways that some critics say it has little predictive power, making it a non-science. A rival theory, loop quantum gravity, says that space-time itself must be quantized, or made from finite chunks. Do the maths, and these chunks turn out to be tiny loops of nothingness, which evolve by themselves into a bubble-like geometry known as a spin foam. The spin foam is essentially space-time but written in the language of quantum mechanics.

Loopy space-time tries to do nothing more than is necessary to reconcile quantum theory and general relativity – a slow and steady approach that some see as more likely to lead to reliable progress. And although loops have remained just as theoretical as strings, loop theorists such as Carlo Rovelli at the Centre of Theoretical Physics in Marseilles, France, have generated some testable predictions (see 'How to test loop quantum gravity' below). That said, there is a hitch. Spin foams are too rigid to readily adhere to the spirit of Einstein's universe where space and time squeeze and stretch, depending on who is looking.

For decades, the two camps have barely talked to each other, but with neither strings nor loops managing to defeat the other side, things are changing as theorists start to move between the two. One connection between the theories emerged in 2011, when Norbert Bodendorfer and colleagues at the University of Warsaw in Poland rewrote string theory, together with its supersymmetric particles, in the space-time described by loop quantum gravity. Another came to light three years later. Loop theorists Rudolfo Gambini of the University of the Republic in Montevideo, Uruguay, and Jorge Pullin of Louisiana State University in Baton Rouge, USA, argued that to make relativity and loop quantum gravity fully compatible, you are forced to limit the range of possible particles using a geometry borrowed from string theory.

The holographic principle

Hints of an even stronger connection date back to the early 1990s, when Gerard 't Hooft at Utrecht University and Leonard Susskind of Stanford proposed the holographic principle: that the three-dimensional world in which we live could be a mere projection of flat, two-dimensional processes taking place on the universe's boundary. They worked out a mathematical correspondence between a 3D theory with gravity and a 2D quantum field theory. In this radical idea, the universe we know is like a three-dimensional image created by the flat hologram on the back of a credit card. Since then, the holographic principle has matured into a major research area in string theory. Theorists find that, from the perspective of a boundary, difficult physics begins to make better sense.

Now it seems that the boundary could provide a place for string theory to entwine with loop quantum gravity. In November 2015, loop theorists Valentin Bonzom of Paris University in France and Bianca Dittrich of the Perimeter Institute for Theoretical Physics in Canada identified a calculation of holographic gravity that others had performed in the context of string theory, and found that exactly the same result could be achieved using loop quantum gravity. A month later, Bodendorfer had another breakthrough. Normally, when holographic string theory is used for calculations of gravity, it gets stuck at singularities like black holes. But Bodendorfer showed that his maths could patch over those situations.

Even more fundamental studies are now weighing in. In November 2016, working with two other collaborators, Laurent Freidel, a theorist at the Perimeter Institute in Waterloo, Canada, went back to basics by using general relativity to describe something very simple: a small region of space surrounded by a boundary. It turned out that the variables defining his boundary bore mathematical resemblances to both string theory and loop quantum gravity, despite neither of those theories being a starting point.

Meanwhile, loop theorist Muxin Han at Florida Atlantic University in Boca Raton, USA, teamed up with string theorist

Ling-Yan Hung at Fudan University in Shanghai, China, to try to calculate the probability of one spin foam evolving into another in loop quantum gravity. Like the soap bubbles that billow while running a bath, spin foams evolve of their own accord – a spontaneity that could help to explain the origin of time. Han and Hung mapped their calculation on to a boundary: once more, mathematical features appeared that were strangely reminiscent of string theory.

If these studies are heading in the right direction, the possibility arises that string theory and loop quantum gravity are, on a boundary at least, not competitors at all but one and the same. But what and where is this boundary? Holographic theory began by imagining a boundary at the universe's edge, but today's string and loop theorists are less strict about this requirement, and think of a boundary as being anywhere in space. Their loop–string physics might appear on the finest scales if you took any random slice through space-time.

That might seem arbitrary. But when you think about it, says Rovelli, we impose boundaries all the time, as portals through which to observe the world. To record light, we catch photons striking the flat surface of photographic film, or an electronic detector, or our retinas. This flat picture can be the result of more than one happening in the bulk, in the same way that the shadow of a butterfly can be generated by an actual butterfly, or by linking your thumbs and waving your fingers. String theory and loop quantum gravity may look different to us, then, but they may, in a sense, cast the same shadow.

There is genuine excitement among younger physicists for where a reconciliation between loops and strings could lead. Even if it does not end at a theory of quantum gravity in itself, it could at least point the way.

FIGURE 9.2 String theory may cast the same shadow as loop quantum gravity.

How to test loop quantum gravity

Loop quantum gravity has been riding high of late, with its theorists generating at least two potentially testable predictions:

- **Black holes that bounce** – Carlo Rovelli has pioneered this idea. If space itself is made of discrete loops, there is a point beyond which it cannot be squashed. That leads him to think that black holes might reach a point where they cannot get any more dense, whereupon they would bounce and produce a burst of observable radiation.
- **Blips from the Big Bang** – The Big Bang is another time when space was very compact and the smallest possible grains of space-time would have been important. It may be that their effects can be seen in cosmic microwave background radiation, the afterglow of the Big Bang. With measurements more precise than we have now, we might detect their signature.

10
After the Large Hadron Collider

In the quest to delve deeper into particle physics, can anything improve on the Earth's biggest machine?

Future colliders: a straight road ahead?

What could replace the LHC? The two most likely successors are the International Linear Collider (ILC) (see Figure 10.1) and the Compact Linear Collider (CLIC). Both would smash electrons and positrons together. The ILC proposal is for a 35-kilometre-long straight accelerator with a collision energy of 1 TeV, whereas the CLIC would reach 3 TeV using a less tested technology. To generate the high-frequency electric fields that accelerate the particles, the ILC would use superconducting resonant cavities made of niobium, whereas the plan at CLIC is to generate radio-frequency fields using a parallel beam of electrons.

Even 3 TeV is a lower collision energy than the LHC. But brute force is not everything. The LHC collides protons, each

FIGURE 10.1 An artist's impression of the proposed International Linear Collider

made of three quarks floating in a bag of force-carrying gluons and short-lived quark pairs. That means the headline energy is shared out among the different parts. Their complexity means protons are blunt instruments: when two of them hit, the result is a confusing array of shrapnel.

Electrons into positrons are single point-like particles, making much sharper instruments. They carry the full advertised energy of the machine, whereas at a proton collider, the energy available for making new particles is only the small share carried by two individual quarks or gluons. And while the state of the colliding quarks and gluons in the LHC is a mystery, the exact energy and other properties of each electron and positron will be known in advance. That makes it possible to work out the precise properties of the Higgs and whatever other exotica might be flung out.

The previous record for electron-positron collisions was achieved at LEP, the accelerator that once occupied the 27-kilometre tunnel now being used for the LHC. Electrons and positrons circled LEP, gradually being boosted to energies as high as 100 GeV. To reach higher, a ring is impractical because circling electrons rapidly lose energy by emitting photons in a process called synchrotron radiation. The faster they go, the faster their energy leaks away. That's why physicists are looking at linear accelerators now. Make the path straight and there are no synchrotron losses, so the plan is to have two separate accelerators pointed head to head.

Narrow beam

Focusing the beam raises its own problems. In a ring, you can make the counter-rotating electron and positron beams cross as often as you want, so the particles have many chances to hit each other. In a linear collider, they have only got one chance.

The solution is to focus the beams down to just a few nano-metres across. That way, each electron is running into a dense mass of positrons (and vice versa), so plenty of collisions should take place. It is difficult, but the necessary techniques have been developed at the Stanford Linear Accelerator Center and at the KEK laboratory in Tsukuba, Japan.

The price tag could be $20 billion and funding is unlikely to come soon – especially as the LHC only seems to have found a relatively straightforward Higgs boson. Physicists had hoped for far more new particles from the LHC – a rich haul that a linear collider would have been able to pick through and examine in detail.

Perhaps there is a better option. Muons are particles similar to electrons, but 200 times more massive. This means that they emit much less synchrotron radiation, so they could be accelerated to high energies in a ring more easily than electrons.

There are a few technical hurdles to be overcome before a muon accelerator becomes reality. Muons are produced in the decay of other particles, pions, as a hot, randomly moving gas, and they need to be cooled before they can be focused into a high-intensity beam. But similar techniques are already used to cool antiprotons, and work is being done to surmount these difficulties at CERN and Fermilab.

As the CLIC and ILC battle it out for funding and support, China has proposed its own alternative. In 2014 scientists at the Institute of High Energy Physics in Beijing announced plans to build a particle collider twice the size of the LHC, with an underground ring at more than 50 kilometres long, to smash electrons and positrons, and also a proton–proton collider in the same tunnel. The aim is to have this facility ready by the 2030s.

The next generation in a nutshell

The International Linear Collider (ILC)

- Status: Design blueprint for this experiment published June 2013
- Summary: A 35-kilometre straight accelerator that smashes electrons and positrons
- Cost: $8 billion
- Pros: Cleaner collisions; technology reliable and well understood
- Cons: In some scenarios, the maximum energy may not be sufficient to see all new physics of interest
- Location: Still to be decided, but the most likely is the Kitakami mountains of Japan

The Compact Linear Collider (CLIC)

The CLIC would be a positron and electron linear accelerator like the ILC – and is also yet to be approved – but it would be shorter and have collisions at higher energies. A high-intensity, low-energy drive beam runs parallel to the colliding beams. Power built up in the drive beam is transferred in quick bursts to the main beams.

- Status: Conceptual design report published October 2012.
- Cost: No official estimate.
- Pros: Cleaner collisions; high energy and compact (the ILC would need to be 140 kilometres long to achieve the same energy, so vastly more expensive)
- Cons: R&D for the new technology still under way
- Location: Unknown

Far in the future
Other proposals include the Very Large Hadron Collider, which would have a collision energy of 40 to 200 TeV and would have to be built from scratch. Muon colliders and an LHeC – smashing an electron beam into a proton beam – are also being considered. In 2014 Chinese scientists announced a particle collider twice the size of the LHC, with a 52-km underground ring to smash electrons and positrons. They aim to have this facility ready by 2028.

Small successors

Building a bigger, more powerful machine than the LHC may be impractical for decades to come. What's more, there is little hope of ever reaching the kind of energies that would directly probe unified forces of nature (see Figure 10.2). So what can particle experimenters do now? One option is to go for precision, not power. That's the reasoning behind a slew of experiments examining familiar particles for infinitesimal signs of deviant behaviour – weirdness that may betray the influence of new phenomena.

Electric squeeze

The standard model predicts that electrons and neutrons should be perfectly round. But it is possible that exotic unknown particles could have some subtle effects on these ordinary ones, squishing or stretching them out of their spherical shape. We know that the properties of ordinary particles are influenced by the presence of virtual particles nearby, and if some of those particles are heavy enough they might give the electron and neutron an electric dipole moment (EDM): a slight separation

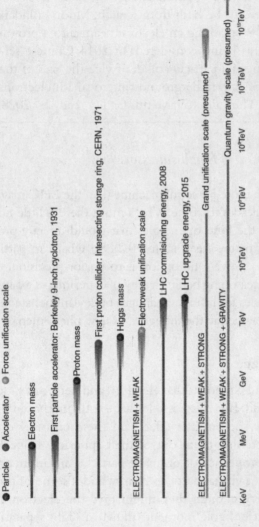

FIGURE 10.2 Particle accelerators have conquered many new frontiers, but are still far short of the energies at which the forces of nature become unified.

between positive and negative charges inside the particle. That makes it an attractive target, particularly for budget-conscious physicists, because experiments searching for EDMs tend to be relatively small-scale and cheap.

The trick is to look very, very closely at these particles' spin. Just as a spinning top wobbles slightly as it slows thanks to the torque applied by gravity, so a particle will wobble in an electric field – that is, if it has an EDM. The trouble is that any such wobble will be incredibly small and therefore hard to spot.

Using supercooled thorium oxide molecules to amplify any deformations, the Advanced Cold Molecule Electron EDM experiment (ACME) at Harvard University has made the most precise measurement of the electron's spherical figure to date. In 2013 it found that any EDM these particles possess must be less than 10^{-30} electron charge metres, a measure of the distance between the positive and negative charges. In other words, if the electron were the size of the Earth, its deviation from perfectly spherical must be equivalent to shaving a sliver less than 10 nanometres wide from the top and slapping it on the bottom.

Meanwhile, the nEDM experiment at Oak Ridge National Lab in Tennessee is probing neutrons. Previous experiments showed that the neutron is round to 1 part per trillion. nEDM is trying to improve that precision by another factor of 100 by embedding the experiment in superfluid helium. This will let the team increase the strength of the electric field acting on the neutrons and slow them down, boosting their chances of seeing any deformation.

Others have suggested trying to spot EDMs in protons, to search for a hypothetical dark-vmatter particle called the axion (see Chapter 7).

These experiments are sensitive to the effects of particles that the LHC may not be capable of seeing directly. At its maximum design energy, the heaviest particles the LHC can find would be about 4 or 5 teraelectronvolts (TeV). By contrast, ACME would already have sensed particles at 7 or 8 TeV, if they existed. The group has already proposed improvements that would push that limit up 40 TeV, and further tweaks could feasibly get them as high as 100 TeV.

Magnetic misbehaviour

The electron's lesser-known cousin, the muon, has been behaving badly for 15 years. Next year we might finally find out what lies behind its disobedience.

Both particles are essentially spinning balls of charge, so they generate a magnetic moment – a north and south pole to you and me. In 1928 Paul Dirac calculated that a quantity related to this magnetic moment, called the g-factor, should be exactly 2 for electrons and muons. But when we measured the electron's magnetic moment in the 1940s, it turned out to be slightly larger: more like 2.002.

We later discovered that this was because virtual particles nudge the electron's magnetic moment off-kilter. The effect is even larger on the muon, which has 207 times the mass of its cousin. That makes it a particularly good candidate for finding new heavy particles, such as those predicted by supersymmetry, as they should add their own little nudges.

Most of the observed differences in magnetic moment come from the influence of known standard-model virtual particles like electrons, positrons and quarks. But in 2001 the E821 experiment at Brookhaven National Laboratory in Upton, New York, showed that the muon's magnetic moment is more deviant, roughly 1 part in 4 billion bigger than it should be, according to the standard model. This implies that there might be some undiscovered heavy particles affecting the muon's magnetic properties. The anomaly was not statistically signifi-cant enough to count as a discovery and the experiment shut down before the team could get what they needed to make the result stick, but a new experiment called Muon g-2 ('gee minus two') gives us another chance to find out.

In 2013 the detector used in E821 – a gigantic ring of super-conducting magnets 15 metres in diameter – was shipped on a barge from New York to Chicago. If all goes well, the team hope that in 2018 they will be able to publish the first result verifying that Brookhaven was right.

The immortal particle

The proton is the bedrock of the atomic nucleus and, accord-ing to the standard model, it is supposed to be absolutely stable: it never disintegrates. But grand unified theories (GUTs) sug-gest that the proton should fall apart eventually.

Super-Kamiokande is a 50,000-tonne tub of water in the mountains of Japan (see Chapter 6). It is best known for its neutrino experiments, but since 1996 the detectors of Super-K have also been watching for a proton to fall apart, an event that would reveal itself through light beams shooting off in opposite directions.

There are several different ways a proton might disintegrate, but one favoured option is decay into a positron and a pion. In 2016 Super-Kamiokande's best limit on the proton lifetime subject to this process was published: a lower limit of 1.6×10^{34} years. That has already ruled out certain types of GUT.

Super-Kamiokande's proposed successor is Hyper-Kamio-kande, which would hold a million tonnes of water. Maybe it will finally see a proton die.

II
Practical particles

Particle physics is not just a quest for abstract truth. It is being used to seek out and kill cancer, to safeguard aircraft wings and to develop better superconductors. One day, neutron science might even stop ketchup from coming out of the bottle in a rush.

What has particle physics ever done for us?

As someone once said, physics is like sex: it may give some practical results but that's not why we do it. This is especially true of particle physics, where the goal is to get a better insight into the nature of the universe by probing matter and forces at the highest energies and the smallest imaginable scales. However, even the alchemists of old knew that the science of turning lead into gold would be pretty useful (i.e. profitable), and likewise it would be silly for particle physicists to completely ignore the impact that their tools and discoveries could have on our everyday lives.

Take particle accelerators. Initially, their development was motivated by the desire to recreate cosmic rays in controlled laboratory conditions – bringing theorists and experimentalists back down to earth. But, perhaps unsurprisingly, the ability to create and control energized beams of charged matter has proven rather handy in other walks of life. For example, electron beams from low-energy accelerators can be used to turn sticky plastic into something more useful. The electrons change the chemistry of the plastic's polymer chains by knocking out some of their hydrogen atoms. That lets the chains join up with each other in a process called crosslinking. The material you end up with is a much stronger plastic that is used to make things like shrink-wrap and electrical insulation.

We also use particle accelerators to generate X-rays for airport security scanners, for irradiating food to keep it fresh, and for sterilizing medical equipment. In the microchip industry, accelerators are used to implant ions in silicon to create more intricate components. And accelerators might yet be used for alchemy – not to turn lead into gold (which is technically possible but massively uneconomical) but instead to de-activate nuclear waste. With a neutron-driven element transmuter, the idea is that you bombard unwanted radioactive atoms with neutrons and they'll change into safer isotopes. This has been shown to work in principle, but still needs refinement to be used at the scales required by the nuclear industry.

The right medicine

Perhaps one of the most exciting developments on the accelerator front is in medical physics: the treatment of cancers with hadron therapy. X-rays have a long history of being used to kill cancer tumours, but the problem is that they also kill every thing else in their way. If you use high-energy protons or carbon ions instead, you can target much more specific areas. Because of the way protons and carbon ions lose their energy as they pass through living tissue, you can get them to deposit their killer blow at a particular depth. This makes hadron therapy a powerful tool for treating tumours in hard-to-reach or delicate areas like the brain or the eye. The reduced radiation dosage required also makes it better for children and other vulnerable patients. It's a beautiful, effective technology born of particle physics. The challenge now is to make the necessary accelerator small enough and cheap enough to install in every hospital that needs one, much like MRI scanners and other such world-changing devices.

One of particle physics' more exotic discoveries – antimatter – is also routinely used in medicine, in positron emission tomography (PET) scans, which harness the power of antimatter to obtain images of what is going on inside the body. To perform a PET scan, you take a substance such as fluorodeoxyglucose that contains radioactive fluorine-18 (which, by the way, is made in a particle accelerator). It gives off a positron, which annihilates with a nearby electron to produce two gamma rays. You can track where those gamma rays appear to pinpoint where that interaction happened, which allows you to produce scans of the brain and other parts of the body that use glucose.

Particle detectors are another story. Since the turn of the twentieth century, the ability to peer into the subatomic world has depended on capturing the telltale trails of the particles as they pass through some sort of sensitive medium. C. T. R. Wilson's cloud chambers used supersaturated water vapour, while Cecil Powell's balloon experiments of the 1940s used a special photochemical emulsion. But setting up these media, and then taking and developing photographs of the particle tracks, was very tiresome. Older readers may remember having to take the film from their cameras to the chemist to get it developed before being able to bore friends and neighbours with their holiday snaps. Carl Anderson had to develop and sort through over 1,300 cloud chamber photographs in order to discover the positron – the antimatter equivalent of the electron. To be fair, he did get a Nobel Prize for his efforts.

As we all have pretty much moved to digital photography, so has the world of particle physics with the use of silicon detectors. As an ionizing particle passes through a suitably prepared slice of silicon, positive and negative charges are bashed about so that they can be fed directly into the detector electronics. The resultant signals can be processed, stored and analysed almost instantly, with no need for any manual processing or development – which was crucial for dealing with the billions of collisions per second required by 2012's discovery of the Higgs boson at the LHC.

Supercomputers

The explosion in the volumes of data made possible by silicon detectors has meant that particle physicists have had to invent new ways to do computing. Grid computing has allowed hundreds of thousands of computers from around the world to simultaneously crunch the numbers flowing from the LHC. In the 1970s Bent Stumpe created the first tracker ball for the CERN control room using a bowling ball, as well as designing one of the first computer touchscreens. And this is quite apart from the World Wide Web invented by Tim Berners-Lee.

This digital revolution has, as you might expect, led to previously unimaginable applications of particle detectors. One such spin-off technology is the Medipix hybrid silicon pixel detector. Compared with the cathedral-sized experiments of the LHC, it is tiny: a grid of 256 × 256 pixels on a silicon wafer about the size of your thumb. But, thanks to some clever microchip engineering, it can be used to detect, count and measure the energy of single photons with micrometre resolution in real time.

This has some huge implications for medical imaging. For example, by looking at photons of different energies as they pass through your subject, you can produce colour X-ray images, opening up a whole new level of diagnostic information to clinicians. And it does not just work for photons. Any ionizing radiation can be detected and visualized instantly, like the alpha and beta radiation from radioactive material, or indeed the protons and carbon ions used in hadron therapy.

Particle detectors are also going to space. One of the main challenges of interplanetary space travel is exposure to dangerous radiation once we are beyond the protective reach of the Earth's coddling magnetic field. If we can measure it and understand it, we should be able to engineer solutions to protect the next generation of explorers as they head towards Mars or Proxima Centauri, or wherever our next frontier lies. There are now five Medipix devices plugged into NASA laptops on the International Space Station, measuring the astronauts' radiation environment. The detectors are also being used by NASA's Orion project, a spacecraft designed to carry astronauts into deep space (see Figure 11.1). So technology developed for particle physics might yet have an impact on our future as a species, as we seek a new place in the universe to call home.

So what's next? Are we going to have Higgs boson-powered warp drives, magnetic monopole monorail systems and micro black hole waste disposal units? Who knows? We don't do science to find applications for these things but, as we have seen, if you have enough people working together in one place – be it CERN, FermiLab or wherever – they are going to come up with some great things.

FIGURE 11.1 Particle detectors have been taken into space aboard NASA's *Orion* spacecraft.

Neutrons at work

After James Chadwick identified the neutron in the 1930s, scientists discovered the neutron's role in converting one chemical element into another, and learned how nuclear reactions and radioactive decay can produce neutrons in large quantity. That led to the discovery of nuclear chain reactions and took physics into the unprecedented territories of nuclear power and nuclear weapons.

But there is another side to the neutron story. Neutrons have become a powerful research tool that can reveal the structure of matter. Today, neutron science touches everything from the next generation of computers to the structure of viruses.

As with all quantum particles, neutrons can behave like waves. So when they encounter obstacles comparable in size to their wavelength, they scatter along well-defined angles, much as water waves diffract around a rock. By analysing the scattering patterns, we can work out the structure of materials that neutrons have passed through.

With the advent of nuclear reactors in the 1940s it became possible to produce high volumes of neutrons. This opened the door to in-depth investigations of the structure of materials. The field really took off in the 1960s, when research reactors were optimized for such experiments.

Intense neutron source

There are now more than 20 active neutron science facilities, which come in two forms. Research reactors such as the High-Flux Reactor at the Laue–Langevin Institute (ILL) in Grenoble, France, use nuclear fission to produce a steady, reliable source of neutrons. The ILL operates the most intense neutron source in the world, feeding beams of neutrons to 40 different instruments. Meanwhile, the ISIS source at the Rutherford Appleton Laboratory in Didcot, UK, accelerates protons into heavy metal targets, prompting the emission of neutrons.

Facilities such as the ILL can produce neutrons over a vast range of energies, and this translates into a wide range of wavelengths. Thermal neutrons, which whizz around at a couple of kilometres per second, have short wavelengths and can be used to study atomic structures less than a nanometre across. Cold neutrons, which move at a tenth of the speed, have long wavelengths and can be used to study molecular structures at the micro scale.

We can now use neutrons to peer inside all manner of materials. They penetrate deep into matter because they have no charge, so they are not deflected by the electric charges of ions. Their spin gives them a small magnetic field that allows them to interact with electron spins, so neutrons are good for understanding the structure and dynamics of magnetic materials. But they interact mainly with the nucleus, through the strong force. That makes neutrons particularly good at identifying the position of light atoms such as hydrogen, oxygen and carbon, because there is a strong contrast between the masses of their nuclei.

Many everyday objects, including tools, clothes, food and health products are made of long chains of hydrocarbons containing these light elements, and neutrons are the best, too, for unravelling their complex structures. In 2012 a team at the ILL and the University of Bristol, UK, used neutrons to find out whether they could make soap magnetic by incorporating iron into the hydrocarbon chains. Such a soap could be manipulated in a magnetic field, improving water treatment and environmental clean-up.

Go with the flow

Other familiar fluids, such as face creams, shampoos and sauces, flow in unusual ways due to the behaviour of their long chain-like molecules. One phenomenon called shear thinning causes a liquid to become much more runny when it is stirred or shaken. It's a process familiar to anyone who has tried to get tomato ketchup out of a glass bottle, only to end up covering the entire plate. To try to find out why this happens, in 2005 scientists fired a beam of neutrons at various liquids as they applied forces to them. The neutrons revealed the molecular orientation, while a device called a rheometer measured the liquid flow. The results established a clear relationship between viscosity and the orientation of the chains, a finding that could help industry predict and tailor how products will leave the bottle or pool in your hand.

Proteins, viruses and cell membranes are naturally rich in light elements, and biologists work alongside neutron scientists to decipher these biological structures and how they carry out their functions. One technique is deuteration. Some or all of the hydrogen atoms in a sample are replaced with deuterium, a heavy isotope of hydrogen that has a neutron in addition to its single proton. Neutron scattering is so sensitive to light elements that it can tell the two isotopes apart. The sample is then contrasted with undeuterated versions to pick out the location of hydrogen atoms in biochemical reactions.

Another area to benefit is the process of introducing foreign DNA into host cells, used in gene therapy and in the genetic modification of crops. Many potential agents for the process have been tested using neutron scattering, including the viruses used to inject DNA into cells.

Neutrons have also provided insight into the transport of cholesterol within cell walls. Cholesterol surrounds every cell, helping to carry signals around the body and assisting in the production of hormones. Maintaining the correct levels of cholesterol by redistributing it between and within the cells is vitally important, with abnormalities linked to Alzheimer's disease and cardiovascular disorders. In recent years, neutrons have illuminated these processes, revealing how cells achieve the right equilibrium and what causes the system to break down.

New drugs

Neutrons are being used not only to aid diagnosis but also to create new drugs. Radiopharmaceuticals are one of the best ways to treat certain tumours. They deliver a radioactive isotope to cancer cells and kill it with a dose of radiation. But the radiopharmaceuticals being used today are the ones that are most readily available rather than the ones with the best properties, and they can cause unnecessary damage to surrounding healthy tissue. Research reactors are now being used to produce new radioisotopes, such as lutetium-177 and terbium-161. In 2015 a team from ILL, the Technical University of Munich in Germany and the Paul Scherrer Institute in Villigen, Switzerland, demonstrated a new technique for producing terbium isotopes in large quantities by irradiating samples of gadolinium with neutrons. Gadolinium-160 absorbs a neutron to produce a heavier isotope that undergoes beta decay and transforms into terbium-161. Terbium-161 emits just enough gamma radiation to help trace the radioisotope's movement around the body, and it emits low-energy electrons that can destroy cancer cells without damaging too much surrounding tissue. It also has a half-life of about one week, long enough for it to be sent to hospitals and short enough not to pose a long-term nuclear waste problem.

The magnetic properties of neutrons are being used to study high-temperature superconductors, which carry a current without any voltage. Researchers are investigating tiny spontaneous loops of current and alternating patterns of spin found within these materials, phenomena thought to play a role in allowing electrons to pair up and move around unimpeded. If we can unlock the secrets of existing high-temperature superconductors, we might be able to make materials that conduct electricity without resistance at room temperature.

These versatile particles can also solve problems with structures that you really don't want to fail, such as aircraft wings, railway tracks and turbine blades. The properties and performance of these materials are largely determined by their nanoscale structure, which is much too small to be examined with ordinary light microscopy. With their shorter wavelengths, neutrons provide a new kind of microscope to understand how stress affects these materials and how their properties can be improved for everyday use.

The hunt for monopoles

Finally, there is the case of the missing monopole. Physicists have searched for magnetic monopoles everywhere, from the debris of high-energy particle collisions to rocks from outer space (see Chapter 8). But neutron experiments have already uncovered a kind of monopole. In 2009 two independent groups found evidence for the phenomenon by firing neutrons at artificial materials called spin ices. The spins of particles in these materials can arrange themselves to create north and south magnetic poles that drift apart independently, analogous to monopoles. One day, these pseudo-monopoles could be used as a form of computer memory far more compact than anything available today.

Neutron futures

Plans are being made for the world's largest neutron science facility, the European Spallation Source (ESS) in Lund, Sweden. It is designed around a linear accelerator that will accelerate protons and collide them with a heavy metal target to release intense pulses of neutrons. These will be guided through beam lines to experimental stations.

The ESS might become the leading light of neutron science, but it certainly won't make the other facilities redundant, because there is so much science to be done in drug discovery, materials science, renewable energy, fundamental physics and biochemistry. Eighty years might have passed since the discovery of the neutron, but we have still only scratched the surface of its potential to revolutionize our world.

Conclusion

As you may have gathered by now, particle physics is more than a dry exercise in cataloguing tiny things. It answers, or is trying to answer, a host of questions about our existence. Many of them are about where we come from in a very literal sense. The familiar, complex matter that makes up our own bodies and almost everything around us depends on a slew of extraordinary events and circumstances dictated by the precise behaviour of particles.

In the past few years, we have found at least partial answers to some of the big mysteries of this field. The precise calculations of lattice quantum chromodynamics show why it is that protons are a little more massive than neutrons. If it were otherwise, there would be no chemical elements. The discovery of the Higgs boson in 2012 confirms why quarks and leptons have mass. Without it, they would all travel at the speed of light and be impossible to bind together into structured matter.

But far more questions remain beyond our present powers. Why did matter particles predominate over antimatter in the early universe? If they hadn't, there would be nothing but a bland sea of radiation. What is dark matter? Without it we would have no stars and galaxies. What sets the energy of empty space? Too high, and the universe is shredded. Could a lightweight Higgs boson make the cosmos unstable? Why does our space-time have three dimensions of space and one of time? What dictates the strengths of all the forces? What kind of particle shenanigans might have caused the violence of inflation? What sparked the Big Bang?

Whatever supersedes the standard model to explain how the basic building blocks of matter interact – be it strings, loops or something stranger – has its work cut out.

Fifty ideas

This section helps you to explore the subject in greater depth, with more than just the usual reading list.

Seven tours and activities

1 **CERN**, the world's biggest physics laboratory and home of the famous LHC particle accelerator, runs guided tours: https://visit.cern/tours. If a trip to Geneva is not feasible, visit via virtual reality here: http://petermccready.com/

2 **The SLAC linear accelerator** in California, said to be the world's straightest object, is worth marvelling at if the LHC is just too curved for your taste: https://www6.slac.stanford.edu/community/public-tours

3 **Westminster Abbey**, London, UK, has a monument to Paul Dirac bearing his eponymous equation, which implied the existence of antimatter.

4 **Rotterdam's Natural History Museum** displays the remains of one of the beech martens (also known as stone martens) electrocuted at the LHC in 2016, causing the particle accelerator to be temporarily shut down.

5 **The Cavendish Laboratory**, Cambridge, UK, was the site of many early discoveries in particle physics, including arguably the earliest, when J.J. Thomson identified the electron using a cathode-ray tube. They have a museum: http://www.phy.cam.ac.uk/outreach/museum. And you can see a short film about that fateful cathode-ray tube here: https://www.newscientist.com/article/2098394-the-tube-that-kicked-off-particle-physics/

6 **Experiment with cathode rays.** You don't need to be a Cambridge scientist to do a cathode-ray experiment. Find an old cathode-ray-tube TV and get a strong magnet. Place the magnet near the screen to bend the particle beam and distort the picture.

7 **Play with neutrinos (and possibly dark matter too).**
Stand, sit or lie anywhere, and remember that a few trillion neutrinos generated in the core of the Sun are passing through your body every second. There may well be
exotic dark-matter particles drifting through you, too.

Eleven quotations

1 'The chances of a neutrino actually hitting something
as it travels through all this howling emptiness are
roughly comparable to that of dropping a ball bearing
at random from a cruising 747 and hitting, say, an egg
sandwich.' Author Douglas Adams (1952–2001)

2 'It is the fact that the electrons cannot all get on top of
each other that makes tables and everything else solid.'
Physicist Richard Feynman (1918–88)

3 'Spreading out the particle into a string is a step in
the direction of making everything we're familiar with
fuzzy. You enter a completely new world where things
aren't at all what you're used to.' M-theory pioneer
Edward Witten (1951–)

4 'I think that this discovery of antimatter was perhaps
the biggest jump of all the big jumps in physics in our
century.' Werner Heisenberg (1901–76), one of the
founding fathers of quantum mechanics

5 'I have committed the ultimate sin, I have predicted
the existence of a particle that can never be observed.'
Quantum pioneer Wolfgang Pauli (1900–58), speaking
of the neutrino

6 'If I could remember the names of all these particles, I'd be a botanist.' Enrico Fermi (1901–54), the Italian physicist who built the first nuclear reactor

7 'I think I might find the universal principles of string theory most elegant – if I only knew what they were.' Leonard Susskind (1940–), one of the fathers of string theory

8 'The electrons perform stately waltzes, weave curvaceous tangos, jitter in spasmodic quicksteps and rock to frenetic rhythms. They are waves dancing to a choreography composed differently for each kind of atom.' British astronomer and writer Edward Harrison (1919–2007), in his 1985 book *Masks of the Universe*

9 'I believe there are 15 747 724 136 275 002 577 605 653 961 181 555 468 044 717 914 527 116 709 366 231 425 076 185 631 031 296 protons in the universe and the same number of electrons.' British astronomer Arthur Eddington (1882–1944)

10 'From the theoretical point of view one would think that monopoles should exist, because of the prettiness of the mathematics. Many attempts to find them have been made, but all have been unsuccessful. One should conclude that pretty mathematics by itself is not an adequate reason for nature to have made use of a theory.' Physicist Paul Dirac (1902–84), who predicted antimatter

11 'All science is either physics or stamp collecting.' Father of nuclear physics Ernest Rutherford (1871–1937)

Six rhymes, jokes and anecdotes

1 A proton walks through a bar. 'I've lost my electron,' it says.
'Are you sure?' says the barman.
'I'm positive.'

2 A Higgs boson walks into a church. 'Hey, you can't have mass without me.'

3 A photon is checking into a hotel, and the doorman asks whether he needs any help with his luggage.
'No, I'm travelling light.'

4 J. J. Thomson was famously absent-minded. One day, persuaded by his colleagues that his only pair of trousers had become too baggy to wear, Thomson bought a new pair on his way home to lunch and changed into them before returning to work. When his wife came home she found the old pair and sent a worried message to the lab, convinced that her husband had gone off in nothing but his underpants, or worse.

5 'As Noether most keenly observed
(And for which much acclaim is deserved),
We can easily see
That for each symmetry,
A quantity must be conserved.' (by David Morin)

6 Wolfgang Pauli, who proposed the existence of neutrinos, was a stickler for scientific rigour. He called some papers and hypotheses 'completely wrong', although that was not the worst epithet: in one famous case, he described a theory as 'not even wrong', because it was untestable.

Four inspired improvisations

1 Cooling the LHC causes the accelerator ring to contract, so individual sections are connected by bellows. After thousands of these components had been put in place, it became clear that some of them were buckling out and obstructing the beam line. But which ones? Someone came up with the idea of putting a ping-pong ball in the tube, where it could coast through the vacuum at a stately couple of metres per second until it hit an obstruction. The ball could be heard clipping off obstacles as it passed through – by a man following its progress on a bicycle. This technique of 'passing the ball' is still used to check that the LHC beam line is clear.

2 Soluble aspirin tablets were used to check for water leaks in the Tevatron accelerator at Fermilab. Each pill held open a switch. During a leak, it would dissolve under the drip, causing the switch to trip, which cut off the electrical supply before damage could occur.

3 The Cryogenic Dark Matter Search detector in Minnesota must be kept less than a degree above absolute zero, so it is insulated with a series of progressively colder layers towards its core. To avoid any contact between the layers, which could cause unwanted warming, stray wires are tied down with dental floss.

4 Gelatine-like konnyaku noodles come in handy for researchers at the high-energy accelerator research organization KEK in Tsukuba, Japan. The noodles' sticky properties mean that they can be used as a test seal for the vacuum needed to make muon detectors.

Nine facts and factinos

1 The fireball created by particle collisions at the LHC can reach several trillion degrees (Celsius, Fahrenheit, Kelvin or Reaumur).

2 The J/Psi meson, whose discovery confirmed the quark model, has its awkward name because it was found independently by two groups, one naming it J, the other Psi.

3 The first cyclotron particle accelerator build by Ernest Lawrence at Berkeley was only about 10 centimetres across.

4 While CERN is known for inadvertently electrocuting unfortunate martens, Fermilab has had its own strained relationship with animals. Its accelerators have been shut down by a cat (tail interrupting a safety beam) and a racoon (chewing a power cable, like the LHC's martens), while other problems have been caused by mice, snakes, geese and deer. Digging muskrats once managed to drain a pond used for cooling water, shutting down the Main Ring accelerator for a day.

5 A happier tale is of Felicia the ferret, employed by Fermilab in the 1970s to clean a long stretch of vacuum pipe.

6 Frank Wilzcek named axions after a brand of laundry detergent.

7 At Cambridge, students graduating with first-class honours in mathematics are known as wranglers; the top student is called senior wrangler. Some great physicists just missed out on this romantic appellation, including

second (not senior) wranglers J. J. Thomson, discoverer of the electron, and James Clerk Maxwell, the founder of electromagnetic theory.

8 UFOs caused problems for the LHC in 2011. Clouds of electrons created by ionized gas in the beam chamber and microscopic dust particles, known as unidentified falling objects, interrupted the beams and made it harder to get the LHC running consistently.

9 J. J. Thomson received a Nobel Prize for discovering that the electron is a particle, while his son George received a Nobel Prize for proving that the electron is a wave. But they are not the only father-and-son prize-winners: five others have been awarded, including to quantum pioneer Niels Bohr and his son Aage.

Three cultural references

1 Antimatter is a popular trope in science fiction, featuring, for example, as the fuel of the Starship *Enterprise* in *Star Trek*.

2 More implausibly, antimatter was used as a plot device in the film *Angels & Demons*, where the Illuminati steal antimatter from the LHC in order to blow up the Vatican.

3 Unlicensed particle accelerators power the proton packs in the film *Ghostbusters*. As their inventor explains, when using these devices it is unwise to cross the streams. By contrast, at the LHC it is necessary.

Ten sites and books for further reading

1 For a very short introduction to particle physics, you probably can't do better than *Particle Physics: A Very Short Introduction* by Frank Close (2004).

2 *Seven Brief Lessons on Physics* (2014), a 'lucid, enchanting' short book by Carlo Rovelli, was a bestseller in Italy, then around the world – for good reason.

3 For delving deeper into the mysteries of particle physics, a good textbook is *Introduction to Elementary Particles* by David Griffiths (2008).

4 For an account of the hunt for the Higgs (and why it matters), try *The Particle at the End of the Universe* by Sean Carroll (2012).

5 Another account of the Higgs discovery is *Massive: The Hunt for the God Particle* by Ian Sample (2010).

6 For an excellent physics travelogue, try Anil Ananthaswamy's *The Edge of Physics: A Journey to Earth's Extremes to Unlock the Secrets of the Universe* (2003).

7 For an exploration of one of the most eccentric physicists, get hold of Graham Farmelo's *The Strangest Man: The Hidden Life of Paul Dirac, Quantum Genius* (2009).

8 To find out more about the unappreciated talent of Emmy Noether and the greatest physics theorem you've probably never heard of, try Dave Goldberg's *The Universe in the Rearview Mirror: How Hidden Symmetries Shape Reality* (2013).

9 If you prefer your information in cunningly folded form, get hold of *Voyage to the Heart of Matter* by Anton Radevsky and Emma Sanders (2009), a pop-up book about CERN's ATLAS detector.

10 And if you'd like to donate some spare computing power to the cause of particle physics, visit LHC@home.

Glossary

ALICE Standing for 'A Large Ion Collider Experiment', this is one of the seven detector experiments at the Large Hadron Collider. It is designed to study the physics of the strong force at extreme energy densities.

Antimatter This is a substance made of antiparticles. Every particle has an antiparticle partner with the same mass but the opposite electric charge. For example, the electron has the positively charged anti-electron, or positron.

ATLAS This is one of two general-purpose detectors at the Large Hadron Collider. It investigates a wide range of physics, from the search for the Higgs boson to particles that could make up dark matter. It has the same scientific goals as the CMS experiment (see below) but uses a different design.

Big Bang According to the Big Bang Theory – our best explanation for why space is expanding – our whole universe exploded from a superhot microscopic region about 13.8 billion years ago.

Boson A particle that carries nature's forces. One of two classes of particle (the other being fermions), according to quantum mechanics, differentiated by a property called spin. Bosons have whole-number spin.

CERN The acronym for the European Organization for Nuclear Research, located astride the French–Swiss border near Geneva.

CMS The Compact Muon Solenoid is a general-purpose detector at the Large Hadron Collider. It uses a huge solenoid magnet to bend the paths of particles from collisions in the LHC.

Dark energy This is thought to dominate the cosmos, making up roughly 68 per cent of everything there is and causing the universe to grow at an ever-increasing pace.

Dark matter A mysterious form of matter that comprises about 27 per cent of everything there is, far outweighing ordinary matter and acting as a gravitational glue to form stars and galaxies.

Electromagnetic force The interaction between electrically charged particles. Electromagnetic interactions are one of the four fundamental interactions (along with gravitational, strong nuclear and weak nuclear).

Electron A negatively charged subatomic particle.

Elementary particle Also known as fundamental particles, these cannot be broken down into constituent parts.

Fermion A particle whose spin is some multiple of a half, such as an electron or a proton.

Flavour The name that scientists use to describe different versions of the same type of particle.

Fundamental particle See elementary particle.

General relativity Einstein's 1915 theory combines the ideas of special relativity and the principle of equivalence into a theory of gravity. Objects bend space-time, making things accelerate towards them.

Gluons These massless particles carry the force that binds quarks together.

Graviton A hypothetical particle that would transmit the force of gravity in quantum theory.

Gravity The weakest of the four known forces of nature, and the only one not explained by the standard model. It seems strong on cosmic scales because it has a long range and is always attractive.

Hadron A subatomic particle made of quarks and anti-quarks, held together by the strong force. The best known hadron is the proton.

Higgs boson A fundamental particle of the standard model. Other particles gain mass through interactions with its associated field. First hypothesized in the 1960, the Higgs boson was finally detected in 2012.

Higgs field Permeates all space and interacts with different particles with varying strengths. Particles that interact more strongly appear heavier. Some particles, such as photons, do not interact with the field at all and remain massless.

Leptons A family of elementary particles. The electron, the muon and the tau are the charged members of the lepton family, and the three types of neutrinos are their uncharged partners.

LHC The Large Hadron Collider, the world's biggest particle accelerator, located at CERN near Geneva.

LHCb The Large Hadron Collider beauty experiment investigates the slight differences between matter and antimatter by studying a particle called the beauty quark, or bottom quark.

Linac An abbreviation for a linear particle accelerator.

Loop quantum gravity LQG attempts to combine general relativity with quantum mechanics. It says that space-time must fit in with the quantum picture: it must come in finite chunks rather than being continuous. These chunks turn out to be tiny loops.

Monopole An isolated magnetic pole. Theorized to exist but not spotted in the wild yet.

Muon This lepton particle has a mass around 200 times greater than the electron.

Neutralino A particle predicted by the theory of supersymmetry, not yet seen in any experiments.

Neutrino These are chargeless, near-massless particles of the standard model. They come in three distinct flavours – electron, mu and tau.

Neutron A subatomic particle with a similar mass to a proton but with no electric charge.

Nucleon A proton or neutron – constituents of the atomic nucleus.

Particle accelerator A machine that accelerates particles such as electrons or protons to very high energies. Magnets are used to focus and steer beams of these particles and they can be made to collide.

Pentaquark A particle consisting of five quarks predicted in the 1960s, and finally detected in 2015.

Photon A massless particle that represents the smallest unit of electromagnetic radiation, or light.

Proton A positively charged subatomic particle found in atomic nuclei. The proton is made up of three quarks.

Quantum chromodynamics QCD is the theory that describes the strong force, which hold quarks together.

Quantum electrodynamics QED is the quantum theory describing the electromagnetic interaction of charged particles.

Quantum field theory This is a framework for constructing quantum mechanical models of subatomic particles.

Quantum mechanics The laws explaining physics at the atomic and subatomic level, where particles move like waves, may be in several states at once and can have shared states connecting them across time and space.

Quarks These building blocks of matter combine to form composite particles called hadrons, the most stable of which are protons and neutrons. There are six types (or flavours) of quark: up, down, strange, charm, bottom and top.

Special relativity Motion, distance and time are relative, according to Einstein's 1905 theory – all because the speed of light is constant.

Spin A particle's intrinsic angular momentum.

Standard model of particle physics This covers the workings of three of the four forces of nature (the electroweak, strong and weak forces, but not gravity). It describes how particles of matter – fermions – feel forces and interact through the exchange of other particles called bosons.

String theory The theory that all particles are manifestations of more fundamental vibrating strings.

Strong nuclear force (or strong force/interaction) This is one of the four fundamental forces of nature. It is the force between protons and neutrons, and between the individual quarks that make them.

Supersymmetry (SUSY) An extension to the standard model that says every particle has a heavier 'superpartner' with slightly different properties.

Theory of everything The all-encompassing yet elusive theory of physics that unifies quantum mechanics and general relativity, and can describe all the forces of nature in a single framework.

W boson This fundamental particle, together with the Z boson, is responsible for the weak force. It was discovered in 1983.

Weak nuclear force (also known as the weak force or weak interaction) One of the four fundamental forces, it is about 10,000 times weaker than the strong force. It is responsible for radioactive decay.

Z boson This is an elementary particle with no electrical charge. The Z boson carries the weak force and was discovered in 1983, just like its electrically charged cousin, the W boson.

Picture credits

Chapter 1

1.1 © Universal History Archive / Universal Images Group / REX / Shutterstock; 1.2 © Prof. Peter Fowler/Science Photo Library; 1.4 © 1998–2017 CERN; 1.5 © John Birdsall/REX/ Shutterstock

Chapter 2

2.4 © Science Photo Library; 2.6 © Design Pics Inc/REX/ Shutterstock

Chapter 3

3.1 © REX/Shutterstock; 3.5 © Callum Bennetts/REX/ Shutterstock

Chapter 4

4.1 © Lucasfilm/20th Century Fox/REX/Shutterstock; 4.2 © Daniel Dominguez/CERN

Chapter 5

5.2 © 2011–2017 CERN; 5.3 © Li Wei/Solent News/REX/ Shutterstock

Chapter 6

6.3 © Felipe Pedreros, IceCube/NSF

Chapter 7

7.2 Source: arxiv.org/pdf/1205.6497.pdf

Chapter 8

8.4 *X-ray:* © NASA/CXC/CfA/M. Markevitch et al.; *Lensing map:* © NASA/STScI; ESO WFI; Magellan/U. Arizona/D.Clowe et al.; *Optical:* NASA/STScI; Magellan/U. Arizona/D. Clowe et al.; 8.5 © Xenon

Chapter 9

9.2 © Chameleons Eye/REX/Shutterstock

Chapter 10

10.1 © Rey.Hori/KEK

Chapter 11

11.1 © NASA/*Orion* Spacecraft

Index